T0141842

Advances in Intelligent Systems and Computing

Volume 737

Series editor

Janusz Kacprzyk, Polish Academy of Sciences, Warsaw, Poland
e-mail: kacprzyk@ibspan.waw.pl

About this Series

The series "Advances in Intelligent Systems and Computing" contains publications on theory, applications, and design methods of Intelligent Systems and Intelligent Computing. Virtually all disciplines such as engineering, natural sciences, computer and information science, ICT, economics, business, e-commerce, environment, healthcare, life science are covered. The list of topics spans all the areas of modern intelligent systems and computing.

The publications within "Advances in Intelligent Systems and Computing" are primarily textbooks and proceedings of important conferences, symposia and congresses. They cover significant recent developments in the field, both of a foundational and applicable character. An important characteristic feature of the series is the short publication time and world-wide distribution. This permits a rapid and broad dissemination of research results.

Advisory Board

More information about this series at http://www.springer.com/series/11156

Ajith Abraham · Abdelkrim Haqiq
Azah Kamilah Muda · Niketa Gandhi
Editors

Proceedings of the Ninth International Conference on Soft Computing and Pattern Recognition (SoCPaR 2017)

 Springer

Editors
Ajith Abraham
Machine Intelligence Research Labs
Auburn, WA
USA

Abdelkrim Haqiq
Faculty of Sciences and Techniques
Hassan 1st University
Settat
Morocco

Azah Kamilah Muda
Faculty of Information and Communication
   Technology
Universiti Teknikal Malaysia Melaka
Durian Tunggal, Melaka
Malaysia

Niketa Gandhi
Machine Intelligence Research Labs
Auburn, WA
USA

ISSN 2194-5357   ISSN 2194-5365   (electronic)
Advances in Intelligent Systems and Computing
ISBN 978-3-319-76356-9   ISBN 978-3-319-76357-6   (eBook)
https://doi.org/10.1007/978-3-319-76357-6

Library of Congress Control Number: 2018935894

Printed on acid-free paper

This Springer imprint is published by the registered company Springer International Publishing AG
part of Springer Nature
The registered company address is: Gewerbestrasse 11, 6330 Cham, Switzerland

Preface

Welcome to the Proceedings of the ninth International Conference on Soft Computing and Pattern Recognition (SoCPaR 2017). Conferences were held at Mogador Hotels & Resorts, Marrakech, Morocco, during December 11–13, 2017. SoCPaR 2016 was held in VIT University, Vellore, India, and this year's event is jointly organized by the Machine Intelligence Research Labs (MIR Labs), USA, and Faculty of Sciences and Techniques, Hassan 1st University, Settat, Morocco.

SoCPaR 2017 is organized to bring together worldwide leading researchers and practitioners interested in advancing the state of the art in soft computing and pattern recognition, for exchanging knowledge that encompasses a broad range of disciplines among various distinct communities. It is hoped that researchers and practitioners will bring new prospects for collaboration across disciplines and gain inspiration to facilitate novel breakthroughs. The themes for this conference are thus focused on "Innovating and Inspiring Soft Computing and Intelligent Pattern Recognition."

The themes of the contributions and scientific sessions range from theories to applications, reflecting a wide spectrum of the coverage of soft computing, intelligent systems, computational intelligence, and its applications. SoCPaR 2017 received submissions from over 20 countries, and each paper was reviewed by at least 5 reviewers in a standard peer-review process. Based on the recommendation by 5 independent referees, finally 18 papers were accepted for publication in the proceedings published by Springer, Verlag.

Many people have collaborated and worked hard to produce a successful SoCPaR 2017 conference. First and foremost, we would like to thank all the authors for submitting their papers to the conference, for their presentations and discussions during the conference. Our thanks go to Program Committee members and reviewers, who carried out the most difficult work by carefully evaluating the submitted papers. Our special thanks to Oscar Castillo (Tijuana Institute of

Technology, Tijuana, Mexico) and Alexander Gelbukh (Instituto Politécnico Nacional, Mexico City, Mexico) for the exciting plenary talks.

We express our sincere thanks to special session chairs and organizing committee chairs for helping us to formulate a rich technical program.

Abdelkrim Haqiq
Ajith Abraham
SoCPaR 2017 - General Chairs

SoCPaR 2017 Organization

Honorary Chairs

Ahmed Nejmeddine	President of Hassan 1st University, Settat, Morocco
Houssine Bouayad	Acting Dean of FST, Hassan 1st University, Settat, Morocco

General Chairs

Abdelkrim Haqiq	GREENTIC, FST, Hassan 1st University, Settat, Morocco
Ajith Abraham	MIR Labs, USA

General Co-chairs

Layth Sliman	EFREI, Paris, France
Adel M. Alimi	University of Sfax, Tunisia

PC Co-chairs

Simone Ludwig	North Dakota State University, USA
Antonio J. Tallón-Ballesteros	University of Seville, Spain
Arpad Kelemen	University of Maryland, USA

Advisory Board

Albert Zomaya	The University of Sydney, Australia
Andre Ponce de Leon F. de Carvalho	University of Sao Paulo at Sao Carlos, Brazil
Bruno Apolloni	University of Milano, Italy
Hideyuki Takagi	Kyushu University, Japan
Imre J. Rudas	Óbuda University, Hungary
Janusz Kacprzyk	Polish Academy of Sciences, Poland
Javier Montero	Complutense University of Madrid, Spain
Krzysztof Cios	Virginia Commonwealth University, USA
Mario Koeppen	Kyushu Institute of Technology, Japan
Mohammad Ishak Desa	Universiti Teknikal Malaysia Melaka, Malaysia
Patrick Siarry	Université Paris-Est Créteil, France
Ronald Yager	Iona College, USA
Salah Al-Sharhan	Gulf University of Science and Technology, Kuwait
Sebastian Ventura	University of Cordoba, Spain
Vincenzo Piuri	Università degli Studi di Milano, Italy

Publication Chairs

Azah Kamilah Muda	UTeM, Malaysia
Niketa Gandhi	Machine Intelligence Research Labs, USA

Web Service

Kun Ma	University of Jinan, China

Publicity Chairs

Mohamed Nemiche	Ibn Zohr University, Agadir, Morocco
Ali Wali	University of Sfax, Tunisia
Brahim Ouhbi	ENSAM, Moulay Ismail University, Meknès, Morocco

Organizing Chairs

Jaouad Dabounou	FST, Hassan 1st University, Settat, Morocco
Mohamed Hanini	FST, Hassan 1st University, Settat, Morocco
Mohamed Chakraoui	Multidisciplinary Faculty of Khouribga, Morocco

Organizing Committee

Youmna El Hiss	FST, Hassan 1st University, Settat, Morocco
Ayman Hadri	FST, Hassan 1st University, Settat, Morocco
Amine Maarouf	xHub, Technopark, Casablanca, Morocco
Ahmed Boujnoui	FST, Hassan 1st University, Settat, Morocco
Hamid Taramit	FST, Hassan 1st University, Settat, Morocco
Adnane El Hanjri	FST, Hassan 1st University, Settat, Morocco
El Mehdi Kandoussi	FST, Hassan 1st University, Settat, Morocco

International Program Committee

Abdelkrim Haqiq	Hassan 1st University, Morocco
Abdellah Ezzati	Hassan 1st University, Morocco
Ajith Abraham	Machine Intelligence Research Labs, USA
Alberto Cano	University of Córdoba, Spain
Aswani Cherukuri	VIT University, India
Azah Kamilah Muda	Universiti Teknikal Malaysia Melaka, Malaysia
Brahim Ouhbi	Université Moulay Ismail, Morocco
Cyril de Runz	University of Reims, France
Daoui Cherki	University Sultan Moulay Slimane, Morocco
Dominique Laurent	Université Cergy-Pontoise, France
Dragan Simic	University of Novi Sad, Serbia
Driss Bouzidi	Mohammed V University, Morocco
Eduardo Solteiro Pires	University of Trás-os-Montes and Alto Douro, Portugal
Eiji Uchino	Yamaguchi University, Japan
Elizabeth Goldbarg	Universidade Federal do Rio Grande do Norte, Brazil
Erkan Bostancı	Ankara University, Turkey
Eulalia Szmidt	Systems Research Institute Polish Academy of Sciences, Poland
Francisco De A. T. De Carvalho	Universidade Federal de Pernambuco, Brazil
Francisco Martine	National Institute for Astrophysics, Optics and Electronics, France

Gerard Dreyfus | ESPCI ParisTech, France
Giorgio Fumera | University of Cagliari, Italy
Giuseppina Gini | Politecnico di Milano, Italy
Jolanta Mizera-Pietraszko | Opole University, Poland
José Valente De Oliveira | Universidade do Algarve, Portugal
Katsuhiro Honda | Osaka Prefecture University, Japan
Kelemen Arpad | University of Maryland, USA
Keun Ho Ryuc | Chungbuk National University, South Korea
Konstantinos Parsopoulosc | University of Ioannina, Greece
Korhan Karabulutc | Yaşar Üniversitesi, Turkey
Kyriakos Kritikosc | Foundation for Research and Technology (FORTH) Hellas, Greece

Laurence Amaralc | Universidade Federal de Uberlândia, Brazil
Lee Chang Yongc | Kongju National University, South Korea
Leocadio G. Casadoc | University of Almería, Spain
Liliana Ironic | Consiglio Nazionale delle Ricerche, Italy
Lin Wangc | Jinan University, China
Lubna Gabrallac | Sudan University of Science and Technology, Sudan

Ludwig Simonec | North Dakota State University, USA
Millie Pant | Indian Institute of Technology Roorkee, India
Muhammad Iqbal | Victoria University of Wellington, New Zealand
Nabil Laachfoubi | Hassan 1st University, Morocco
Nadia El Mrabet | LIASD - Université Paris 8, France
Niketa Gandhi | Machine Intelligence Research Labs, USA
Nizar Rokbani | École Nationale d'Ingénieurs de Sfax, Tunisia
Oscar Castillo | Instituto Tecnológico de Tijuana, Mexico
Pranab Muhuri | South Asian University, India
Said El Kafhali | Hassan 1st University, Morocco
Saverio De Vito | ENEA, Italy
Swati V. Shinde | Pimpri Chinchwad College of Engineering, India
Tarun Sharma | Indian Institute of Technology Roorkee, India
Valeriu Beiu | United Arab Emirates University, United Arab Emirates

Varun Kumar Ojha | Swiss Federal Institute of Technology, Switzerland
Vincenzo Piuri | Università degli Studi di Milano, Italy
Virgilijus Sakalauskas | Vilnius University, Lithuania
Wei Chiang Hong | Oriental Institute of Technology, Taiwan
Wen Yang Lin | National University of Kaohsiung, Taiwan
Weng Kin Lai | The Institution of Engineering and Technology, Malaysia

Additional Reviewers

Maria Salvato	University of Maryland, Baltimore, USA
Sandeep Kumar	South Asian University, New Delhi, India
Youssef Baddi	ENSIAS - Mohammed V-Souissi University, Rabat, Morocco
Fabiano Dorça	Federal University of Uberlândia, MG, Brazil
Elena Esposito	University of Lausanne (DEEP-HEC), Switzerland

Contents

An Energy-Efficient MAC Protocol for Mobile Wireless Sensor Network Based on IR-UWB

Anouar Darif[1](✉), Chaibi Hasna[2], and Rachid Saadane[2]

[1] LRIT-GSCM Associated Unit to CNRST (URAC 29) FSR,
Mohammed V-Agdal University, BP 1014, Rabat, Morocco
anouar.darif@gmail.com
[2] SIR2C2S/LASI-EHTP, Hassania School of Public Labors,
Km 7 El, Jadida Road, BP 8108, Casa-Oasis, Casablanca, Morocco
has.chaibi@gmail.com, saadane@ehtp.ac.ma

Abstract. Mobile Wireless Sensor Network (MWSN) owes its name to the presence of mobile sensor nodes within the network. It has recently launched a growing popular class of WSN in which mobility becomes an important area of research for the WSN community. In this type of network the energy efficiency is the key design challenge. For this reason MAC layer protocols for MWSN based on IR-UWB must be energy efficient to exploit the main features of IR-UWB technology implemented in the physical layer and maximize lifetime. In this paper we present and show the good impact in term of energy consumption for an energy-efficient MAC protocol in MWSN based on IR-UWB. This MAC protocol takes advantage of these two key properties by using asynchronous periodic beacon transmissions from each network node and its duty-cycling mode.

We developed our own class MWideMacLayer under MiXiM platform on OMNet++ platform to test and evaluate the performance of WideMac protocol compared to ALOHA and Slotted ALOHA.

Keywords: MWSN · IR-UWB · WideMac · ALOHA · Slotted ALOHA
Energy · PDR

1 Introduction

MWSN is one of the most interesting networking technologies since its ability to use no infrastructure communications, it have been used for many applications, including military sensing, data broadcasting [1], environmental monitoring [2], Intelligent Vehicular Systems [3], multimedia [4], patient monitoring [5], agriculture [6], and industrial automation [7] etc. This kind of networks has not yet achieved widespread deployments, though it has been proven able to meet the requirements of many classes of applications. Mobile wireless sensor nodes have some limitations as lower computing capabilities, smaller memory devices, small bandwidth and very lower battery autonomy; these constraints represent the main challenges in the development or deployment of any solution using MWSN. Energy consumption is a very important design consideration in MWSN, New wireless technologies emerge in the recent few

© Springer International Publishing AG, part of Springer Nature 2018
A. Abraham et al. (Eds.): SoCPaR 2017, AISC 737, pp. 1–12, 2018.
https://doi.org/10.1007/978-3-319-76357-6_1

years, providing large opportunities in terms of low power consumption, high and low rate and are promising for environment monitoring applications. IR-UWB technology is one of these new technologies; it is a promising solution for MWSN due to its various advantages such as its robustness to severe multipath fading even in indoor environments, its potential to provide accurate localization, its low cost and complexity, and low energy consumption [9]. It is necessary to find a very adapt MAC layer protocol to this Technology for keeping his advantages.

The present paper is organized as follows. In Sect. 1 we introduced MWSN. In Sect. 3 we presented the IR-UWB technology. Section 4 presents WideMac protocol. The simulation and its results are presented in Sect. 5; finally, Sect. 6 concludes the paper.

2   MWSN Overview

2.1   MWSN Architectures

MWSN can be categorized by flat, two-tier, or three-tier hierarchical architectures.

Flat: In this case, the network architecture comprises a set of heterogeneous devices that communicate in an ad hoc manner. The devices can be mobile or stationary, but all communicate over the same network. As an example, the basic navigation systems had a flat architecture [9].

Two-tier: This architecture consists of a set of stationary nodes, and a set of mobile nodes. The mobile nodes form an overlay network or act as data mules to help move data through the network. The overlay network can include mobile devices that have greater processing capability, longer communication range, and higher bandwidth. Furthermore, the overlay network density may be such that all nodes are always connected, or the network can become disjoint. When the latter is the case, mobile entities can position themselves in order to re-establish connectivity, ensuring network packets reach their intended destination. The NavMote system takes this approach [10].

Three tier: In this architecture, a set of stationary sensor nodes pass data to a set of mobile devices, which then forward that data to a set of access points. This heterogeneous network is designed to cover wide areas and be compatible with several applications simultaneously. For example, consider a sensor network application that monitors a parking garage for parking space availability. The sensor network (first layer) broadcasts availability updates to compatible mobile devices (second layer), such as cell phones or PDAs that are passing by. In turn, the cell phones forward this availability data to access points (third layer), such as cell towers, and the data are uploaded into a centralized database server. Users wishing to locate an available parking space can then access the database.

2.2   Node Roles

At the node level, mobile wireless sensors can be categorized based on their role within the network:

Mobile Embedded Sensor: Mobile embedded nodes do not control their own movement; rather, their motion is directed by some external force, such as when tethered to an animal [11] or attached to a shipping container [12]. Typical embedded sensors include [13, 14].

Mobile Actuated Sensor: Sensor nodes can also have locomotion capability, which enables them to move throughout a sensing region. With this type of controlled mobility, the deployment specification can be more exact, coverage can be maximized, and specific phenomena can be targeted and followed [15–17].

Data Mule: Oftentimes, the sensors need not be mobile, but they may require a mobile device to collect their data and deliver it to a base station. These types of mobile entities are referred to as data mules. It is generally assumed that data mules can recharge their power source automatically [18].

Access Point: In sparse networks, or when a node drops off the network, mobile nodes can position themselves to maintain network connectivity. In this case, they behave as network access points [19].

3   IR-UWB

IR-UWB is a promising technology to address MWSN constraints. However, existing network simulation tools do not provide a complete MWSN simulation architecture, with the IR-UWB specificities at the Physical (PHY) and the Medium Access Control (MAC) layers. The IR-UWB signal uses pulses baseband a very short period of time of the order of a few hundred picoseconds. These signals have a frequency response of nearly zero hertz to several GHz. According to [20] there is no standardization, the waveform is not limited, but its features are limited by the FCC mask. There are different modulation schemes baseband for IR-UWB [21]. This paper uses the PPM technique for IR-UWB receiver.

3.1   IR-UWB Pulse Position Modulation

Pulse position modulation can be represented as follows at the transmitter:

$$S(t) = \sqrt{E} \sum_j P_0 \left(t - jT_{sym} - \theta_j - b_j T_{sym}/2 \right) \tag{1}$$

Where: E is the pulse energy, $P_0(t)$ is the normalized pulse waveform, T_{sym} is the symbol duration, θ_j is the time-hopping shift for the considered symbol j, b_j is the jth bit value and $T_{sym}/2$ is the time shift for the modulation.

We Considering an AWGN channel of impulse response:

$$h(t) = \lambda \delta(t - \tau) \tag{2}$$

Where λ is the attenuation and τ is the delay, the energy at the receiver is $E_u = \lambda^2 E$. The signal after propagation becomes:

$$r_u(t) = \sqrt{E_u} \sum_j P_0 \left(t - jT_{sym} - \theta_j - b_j T_{sym}/2 - \tau \right) \tag{3}$$

The received signal can be separated into three components: $r_u(t)$, $r_{mai}(t)$ and n(t).

Where: $r_u(t)$ is the transmitted signal from the source transformed by the channel, $r_{mai}(t)$ is the multiple access interference caused by simultaneous transmissions and n(t) is the thermal noise.

The thermal noise is a zero-mean Gaussian random variable of standard deviation $N_0/2$ (where N_0 is the thermal noise given by $N_0 = K_B T$, K_B being the Boltzmann constant and T the absolute temperature).

The multiple access interference can be expressed as follows:

$$r_{mai}(t) = \sum_{n=1}^{N_i} \sqrt{E^{(n)}} \times \sum_j P_0 \left(t - jT_{sym} - \theta_j^{(n)} - \frac{b_j^{(n)} T_{sym}}{2} - \tau^{(n)} \right) \tag{4}$$

Where: N_i is the number of interfering signals, $E^{(n)}$ is the received energy, $\tau^{(n)}$ is the channel delay for the considered signal, $\theta_j^{(n)}$ is the time-hopping shift and $b_j^{(n)}$ is the bit value for the jth symbol of the considered interfering signal.

The correlating received signal $S_{Rx}(t)$ with a correlation mask m(t) effect can be expressed as:

$$Z(x) = \int_{\tau}^{\tau + T_s} S_{Rx}(t) m_x(t - \tau) dt = Z_u + Z_{mai} + Z_n \tag{5}$$

With:

$$m_x(t) = P_0 \left(t - xT_s - \theta_j \right) - P_0 \left(t - xT_s - \theta_j - T_s/2 \right) \tag{6}$$

The signal contribution (Z_u), the thermal noise contribution (Z_n) and multiple access interference (Z_{mai}) are the decision variable Z(x) component.

With Z_n is Gaussian distributed with zero mean and variance:

$$\sigma_n^2 = N_0 (1 - R_0(\varepsilon)) \tag{7}$$

With:

$$R_0(t) = P_0(\varepsilon) P_0(\varepsilon - t) d\varepsilon \tag{8}$$

Considering Z_{mai} is Gaussian distributed with zero mean and variance:

$$\sigma_{mai}^2 = \frac{1}{T_s} \sigma_M^2 \sum_{n=1}^{N_i} E^{(n)} \tag{9}$$

With:

$$\sigma_M^2 = \int_{-\infty}^{+\infty} \left(\int_{-\infty}^{+\infty} P_0(t-\tau)(P_0(t) - P_0(t-\varepsilon)) dt \right)^2 d\tau \qquad (10)$$

3.2   Radio State Machine

The power consumption is derived from the time spent in each of the radio modes, it is important to model these accurately. We use the finite state machine illustrated on Fig. 1, with three steady states Sleep, Rx and Tx, and four transient states SetupRx, SetupTx, SwitchRxTx and SwitchTxRx. The radio can always leave any state (steady or transient) and immediately enter sleep mode. The time spent in a transient state is a constant $T_{TrState}$, the power consumption in each state is P_{State} and the energy cost of a transition from one steady state to another is $E_{TrState}$.

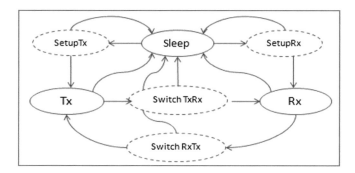

Fig. 1. Detailed radio model including transient states

4   WideMac

4.1   Presentation

WideMac was presented as a novel MAC protocol designed for MWSN using ultra wide band impulse radio transceivers. It makes all nodes periodically (period T_W, identical for all nodes) and asynchronously wake up, transmit a beacon message announcing their availability and listen for transmission attempts during a brief time T_{Listen}.

Figure 2 illustrate a single period structure. It starts with a known and detectable synchronization preamble and is followed by a data sequence which announces the node address and potentially other information, such as a neighbor list or routing table information (for instance, cost of its known path to the sink). A small listening time follows T_{Listen}, during which the node stays in reception mode and that allows it to receive a message [22].

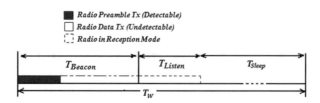

Fig. 2. Detailed view of a WideMac period

The whole period composed of T_{beacon} and T_{Listen} is called T_a (time of activity); and its very small compared to the time window T_W. This period is followed by a long sleeping period T_{Sleep} during which nodes save energy by keeping the radio in its sleeping mode.

When a node has a message to transmit, it first listens to the channel until it receives the beacon message of the destination node. This beacon message contains a backoff exponent value that must be used by all nodes when trying to access this destination. If this value is equal to zero, the source node can transmit immediately. Otherwise, it waits a random backoff time, waits for the destination beacon, and transmits its data packet. Because of the unreliability of the wireless channel, packets are acknowledged. If a packet is not acknowledged, or if the destination beacon was not received a retransmission procedure using the backoff algorithm is initiated, until the maximum number of retransmissions maxTxAttempts is reached.

The details of the backoff algorithm are described in subsection. Figure 3 depict a sender node listening to the channel, ignoring the beacon message of another node, and sending its message to the destination after receiving its beacon. The exchange ends with an acknowledgment message transmitted by the receiver node and addressed to the sender node [23].

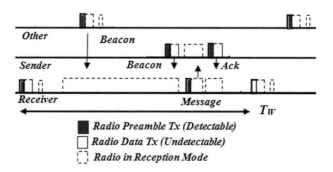

Fig. 3. An initial WideMac ata transmission

4.2   WideMac Backoff Algorithm

The backoff algorithm has a major effect on collision, latency and fairness. WideMac periodic beacons allow the sender nodes to get some information on the channel state at the destination. This can be used to reduce the hidden and exposed terminal problems.

The WideMac transmission procedure works as follows: a candidate sender node first listens for the receiver node's beacon. Once it finds it, it can either immediately attempt transmission (default for lightly loaded networks) or it can start a backoff timer before sending (this is activated by a flag always Backoff in the beacon). In both cases, the sender node waits for an acknowledgment. If it does not arrive, a retransmission procedure begins. The sender node chooses a random time parameterized by the receiver node's Backoff Exponent (BE) which was broadcast in the beacon, using a binary exponential backoff:

$$T_{Backoff} = N_{Backoff} . T_W, \tag{11}$$

$$\text{where } N_{Backoff} \in \left[0, 2^{BE_{Receiver}} - 1\right] \tag{12}$$

The backoff time is thus a function of the wake-up interval T_W and of the channel state at the receiver node, as captured by $BE_{Receiver}$. Such a receiver-based backoff parameterization was also proposed in IR-MAC [24]. The use of a slotted backoff time based on T_W is natural since all candidate sender nodes are synchronized on the receiver node's wake up times: using a fraction of T_W would not change anything as the node would not transmit before receiving the destination beacon. Using an integer multiple of T_W for the unit backoff duration would increase latency and spread the traffic, but this can also be achieved by adapting the value of $BE_{Receiver}$ to the traffic conditions.

4.3   Power Consumption Models

Each normal T_W interval starts with a beacon frame transmission followed by a packet or a beacon reception attempt, during this start a node must enter transmission mode ($E_{SetupTx}$), transmit its beacon ($T_{Beacon}P_{Tx}$), switch to reception mode (E_{SwRxTx}) and attempt a packet reception ($T_{Listen}P_{Rx}$). These costs are regrouped in the beacon energy E_{Beacon}.

$$E_{Beacon} = E_{SetupTx} + T_{Beacon}P_{Tx} + E_{SwTxRx} + T_{Listen}P_{Rx} \tag{13}$$

In addition, during a time L, a node must sometimes transmit a packet E_{Tx} or receive one E_{Rx}, and sleep the rest of the time E_{Sleep}, resulting to the following average power consumption:

$$P_{WideMac} = \frac{1}{T_W} \left(E_{Beacon} + E_{Tx} + E_{Rx} + E_{Sleep}\right) \tag{14}$$

Where:

$$E_{Tx} = K.C_{Tx}(P_{out}).V_B.T_{Tx} \tag{15}$$

$$E_{Rx} = K.C_{Rx}.V_B.T_{Rx} \tag{16}$$

$$E_{Sleep} = C_{Sleep}.V_B.T_{Sleep} \qquad (17)$$

K represents the message length in bytes, P_{out} is the transmission power, C_{Tx}, C_{Rx} and C_{Sleep} represent the current intensities for the three modes, T_{Tx} and T_{Rx} are the time of transmission and reception.

5   Simulations and Results

5.1   OMNet++ and MiXiM Simulation Platform

OMNeT++ is an extensible, modular, component-based C++ simulation library and framework which also includes an integrated development and a graphical runtime environment; it is a discreet events based simulator and it provides a powerful and clear simulation framework.

MiXiM joins and extends several existing simulation frameworks developed for wireless and mobile simulations in OMNeT++. It provides detailed models of the wireless channel, wireless connectivity, mobility models, models for obstacles and many communication protocols especially at the Medium Access Control (MAC) level. Moreover, it provides a user-friendly graphical representation of wireless and mobile networks in OMNeT++, supporting debugging and defining even complex wireless scenarios [25].

5.2   Simulation Parameters

We performed the simulations in the MiXiM 2.1 release framework with the OMNeT++ 4.2 network simulator.

To test and evaluate the performance of WideMac protocol we used PhyLayer-UWBIR class developed under MiXiM platform on OMNet++ as a physical layer. For the MAC layer we developed our own class MWideMacLayer.

We used a grid network, where nodes transmit packets to a Sink node; also we ran several simulations with different nodes numbers and parameters values to evaluate our new protocol.

For the energy consumption we used the following radio power consumption parameters shown in Table 1.

Table 1. Energy parameters

Parameter	Value
P_{Rx}	36.400 mW
P_{Tx}	1.212 mW
P_{Sleep}	0.120 mW
$P_{SetupRx}$	36.400 mW
$P_{steupTx}$	1.212 mW
P_{swTxRx}	36.400 mW
P_{swRxTx}	36.400 mW

For the radio timing we used the parameters shown below in Table 2.

Table 2. Timing parameters

Parameter	Value
$T_{SetupRx}$	0.000103 s
$T_{SetupTx}$	0.000203 s
T_{SwTxRx}	0.000120 s
T_{SwRxTx}	0.000210 s
$T_{RxToSleep}$	0.000031 s
$T_{TxToSleep}$	0.000032 s
Bit rate	0.850000 Mbps

5.3   Results

a. *Energy consumption*

In this section, we present the results obtained using the timing and energy parameters cited in Sect. 5.2. The energy-efficient of WideMac was concretized by the results shown in Figs. 4 and 5. They show that the power consumption of WideMac protocol is remarkably less than the ALOHA and Slotted ALOHA MAC protocols. The good result obtained in the case of WideMac due to the duty-cycling mode uses by this protocol. This mode keeping the radio of wireless communication systems in sleep mode as much as possible.

Figure 4 shows the result obtained by a mobile nodes' number fixed at 40 nodes and varying the data payload size. It shows clearly that the value of power consumption increase with increasing the data payload size due to the required power for sending all

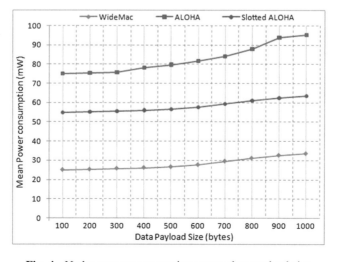

Fig. 4. Nodes power consumption versus data payload size

data packet. The result shown in Fig. 5 is obtained by a data payload fixed at 600 bytes and varying the mobile nodes' number. It shows also the linear dependence between the power consumption and the nodes' number.

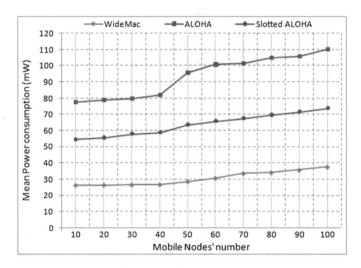

Fig. 5. Nodes power consumption versus mobile nodes' number

b. *Packets Delivery Ratio (PDR)*

To implement a good solution for such a system, the quality of service has to be taken into consideration which explains our study of the Packets delivery ratio parameter. Figure 6 shows a good packets delivery ratio in the WideMac case which reaches

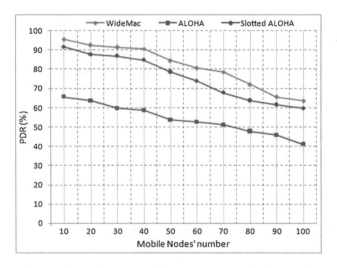

Fig. 6. Packets delivery ratio versus mobile nodes' number

95,56% in the scenario of 10 mobile nodes' number which is equal respectively to 65,47% and 91,67% in the ALOHA and Slotted ALOHA. This figure proves clearly the influence of the mobile nodes' number on this parameter in three cases, because the packets delivery ratio parameter is a direct result of the efficiency of both physical and MAC layers.

6 Conclusion

In this paper we showed the impact and the gain brought by the use of the WideMac protocol for MWSN based on IR-UWB in terms of energy consumption and PDR compared to the ALOHA and Slotted ALOHA. The low energy consumption is the main advantage of the WideMac protocol; it is also very close to an ideal energy consumption model for the IR-UWB based transceivers and gave a good result at this level. This result was achieved thanks to the fact that the network nodes are sleep in the T_{sleep} periods which occupy a wide range in the T_W periods.

We aim, as a future work, to develop a new adapted routing protocol that will be paired with WideMac to largely exploit the IR-UWB features in MWSN.

References

1. Sung, W., Wu, T.-T., Yang, C.-S., Huang, Y.-M.: Reliable data broadcast for Zigbee wireless sensor networks. Int. J. Smart Sensing Intell. Syst. **3**(3) (2010)
2. Jang, W.S., Healy, W.M.: Assessment of performance metrics for use of WSNs in buildings. In: International Symposium on Automation and Robotic in Construction (ISARC 2009), 27–29 June 2009, pp. 570–575 (2009)
3. Mouftah, H.T., Khanafer, M., Guennoun, M.: Wireless sensor network architectures for intelligent vehicular systems. In: Symposium International for Telecommunication Techniques (2010)
4. Suh, C., Mir, Z.H., Ko, Y.-B.: Design and implementation of enhanced IEEE 802.15.4 for supporting multimedia service in wireless sensor networks. Int. J. Comput. Telecommun. Netw. **52**(13), 2568–2581 (2008)
5. Golmie, N., Cypher, D., Rebala, O.: Performance analysis of low rate wireless technologies for medical applications. J. Comput. Commun. **28**(10), 1266–1275 (2005). ISSN 0140-3664
6. Zhoul, H., Chen, X., Liu, X., Yang, J.: Applications of Zigbee wireless technology tomeasurement system in grain storage. In: Computer and Computing Technologies in Agriculture II. IFIP International Federation for Information Processing, vol. 3, pp. 2021–2029 (2009). https://doi.org/10.1007/978-1-4419-0213-952
7. Willig, A.: Recent and emerging topics in wireless industrial communication. IEEE Trans. Industr. Inf. **4**(2), 102–124 (2008)
8. Lecointre, A., Berthe, A., Dragomirescu, D., Plana, R.: Performance evaluation of impluse radio ultra wide band wireless sensor networks. In: Proceedings of the 28th IEEE Conference on Military Communications, MILCOM 2009, pp. 1191–1197 (2009)
9. Amundson, I., Koutsoukos, X., Sallai, J.: Mobile sensor localization and navigation using RF doppler shifts. In: 1st ACM International Workshop on Mobile Entity Localization and Tracking in GPS-Less Environments, MELT 2008 (2008)

10. Fang, L., Antsaklis, P.J., Montestruque, L., Mcmickell, M.B., Lemmon, M., Sun, Y., Fang, H., Koutroulis, I., Haenggi, M., Xie, M., Xie, X.: Design of a wireless assisted pedestrian dead reckoning system – the NavMote experience. IEEE Trans. Instrum. Meas. **54**(6), 2342–2358 (2005)
11. Juang, P., Oki, H., Wang, Y., Martonosi, M., Peh, L., Rubenstein, D.: Energy-efficient computing for wildlife tracking: design tradeoffs and early experiences with ZebraNet. In: Proceedings of ASPLOS-X (2002)
12. Kusy, B., Ledeczi, A., Koutsoukos, X.: Tracking mobile nodes using RF doppler shifts. In: Proceedings of the 5th International Conference on Embedded Networked Sensor Systems, SenSys 2007, pp. 29–42. ACM, New York (2007)
13. Dutta, P., Grimmer, M., Arora, A., Bibyk, S., Culler, D.: Design of a wireless sensor network platform for detecting rare, random, and ephemeral events. In: Proceedings of IPSN/SPOTS, April 2005
14. Polastre, J., Szewczyk, R., Culler, D.: Telos: enabling ultra-low power wireless research. In: Proceedings of IPSN/SPOTS, April 2005
15. Dantu, K., Rahimi, M., Shah, H., Babel, S., Dhariwal, A., Sukhatme, G.S.: Robomote: enabling mobility in sensor networks. In: The Fourth International Symposium on Information Processing in Sensor Networks, IPSN 2005 (2005)
16. Friedman, J., Lee, D.C., Tsigkogiannis, I., Wong, S., Chao, D., Levin, D., Kaisera, W.J., Srivastava, M.B.: Ragobot: a new platform for wireless mobile sensor networks. In: International Conference on Distributed Computing in Sensor Systems, DCOSS 2005 (2005)
17. Bergbreiter, S., Pister, K.S.J.: CotsBots: an off-the-shelf platform for distributed robotics. In: Proceedings of the IEEE/RSJ International Conference on Intelligent Robots and Systems, IROS 2003 (2003)
18. Shah, R., Roy, S., Jain, S., Brunette, W.: Data mules: modeling a three-tier architecture for sparse sensor networks. In: Proceedings of the First IEEE International Workshop on Sensor Network Protocols and Applications (2003)
19. Wang, G., Cao, G., Porta, T., Zhang, W.: Sensor relocation in mobile sensor networks. In: IEEE INFOCOM 2005 (2005)
20. Lazaro, A., Girbau, D., Villarino, R.: Analysis of vital signs monitoring using an IR-UWB radar. Prog. Electromagnet. Res. **100**, 265–284 (2010)
21. Adsul, A.P., Bodhe, S.K.: Performance comparison of BPSK, PPM and PPV modulation based IR-UWB receiver using wide band LNA. Int. J. Comput. Technol. Appl. **3**(4), 1532–1537 (2012)
22. Piguet, D., Decotignie, J.-D., Rousselot, J.: A MAC protocol for micro flying robots coordination. European Community's Seventh Framework Programme (FP7/2007-2013) under grant agreement no 231855 (sFly)
23. Rousselot, J., El-Hoiydi, A., Decotignie, J.-D.: WideMac: simple and efficient medium access for UWB sensor networks. In: IEEE International Conference on Ultra-Wideband (2008)
24. Sun, Y., Gurewitz, O., Johnson, D.B.: RI-MAC: a receiver- initiated asynchronous duty cycle MAC protocol for dynamic traffic loads in wireless sensor networks. In: Proceedings of the 6th ACM Conference on Embedded Network Sensor Systems, SenSys 2008, pp. 1–14. ACM, New York (2008)
25. Köpke, A., Swigulski, M., Wessel, K., Willkomm, D., Klein Haneveld, P.T., Parker, T.E.V., Visser, O.W., Lichte, H.S., Valentin, S.: Simulating wireless and mobile networks in OMNeT++ the MiXiM vision. In: Proceedings of International Workshop on OMNeT++ (co-located with SIMUTools 2008), March 2008

Obstacle Detection Algorithm by Stereoscopic Image Processing

Navigation Assistance for the Blind and Visually Impaired

Adil Elachhab$^{(\boxtimes)}$ and Mohammed Mikou

Laboratory: Analysis of Systems and Information Processing,
Faculty of Sciences and Techniques, University Hassan I,
BP 577, 26000 Settat, Morocco
adil.elachhab@gmail.com, mohammed.mikou@uhp.ac.ma

Abstract. The continued growth in the number of blind or visually impaired people calls for an urgent need to invest all efforts to improve the quality of life for this population. This paper presents a work forming part of the technical study of new aid to the autonomous and secure movement for blind and partially sighted persons. These are intelligent glasses that capture environmental information, via a pair of stereoscopic cameras. In the case where an obstacle is present in the path of visual impairment, it will be detected by the stereoscopic sensor and an obstacle avoidance algorithm incorporated in a laptop computer. The final result, which provides key data to describe its environment, is transmitted to the user via headphones. Under this goal, we have developed a specialized processing algorithm that uses a stereoscopic vision approach and consists to determine the distance to obstacles and their size from two views of the same scene.

The results provided by the algorithm seem to be satisfactory in terms of computing time required and effective operating range.

Keywords: Obstacle detection · Visually impaired · Stereoscopic vision
Subpixel · Dynamic programming · Pyramidal image · Image segmentation
Super-pixels

1 Introduction

A visually impaired encounters a difficulty to reliably map the depth of the environment during its travels to an unknown destination. This limitation of accessibility, therefore, reduces its autonomy. A thorough review of aid systems based on RFID (Radio Frequency Identification) [1, 2], GPS (Global Positioning Systems) [3, 4], and ultrasound [5, 6], shows that despite efforts, these systems do not always provide a safe and independent mobility solution for the visually impaired. It appears necessary to find alternatives to determine the position of the obstacles continuously and in real time, and to extract the information necessary for navigation.

© Springer International Publishing AG, part of Springer Nature 2018
A. Abraham et al. (Eds.): SoCPaR 2017, AISC 737, pp. 13–23, 2018.
https://doi.org/10.1007/978-3-319-76357-6_2

The arrival of stereoscopic vision system has been a source of great hope, because it is one of the most promising tracks in the navigation aid systems supposed to locate with certainty an obstacle and guide a blind or partially sighted person through his path [7, 8]. Stereoscopic vision systems extract depth information from two images acquired by two separate cameras placed at a distance B. The two cameras capture the scene that is then transmitted to a processing unit on which the proposed algorithm is integrated [9, 10]. The result of the processing device is further communicated to the user through an audible voice through earphones. In this context, we propose a new obstacle detection algorithm based on processing a pair of images taken by two synchronized cameras with parallel optical axes. In these conditions, two issues appear: how to successfully recover the objects depth in a scene in order to estimate the distance traveled before touching an obstacle? How can both get this reliably distance information, and within a reasonable time.

In the previous work [11–13], we proposed an algorithm that determines the obstacles position for both large and small obstacle-camera distances, evaluating the disparities between the two stereoscopic images by the algorithm "Block Matching" using the *SAD* criterion (Sum of Absolute Differences). Nevertheless, the results achieved are encouraging in terms of accuracy and reduced computational load, but they are not suitable for real-time applications. In this study, we propose to improve our algorithm by first considering the separation by segmentation the obstacles of the background before image treatment and the integration of advanced optimization approaches, resulting in a fast stereoscopic image processing. We also propose using this algorithm to determine the limit positions for the obstacle location, for which the visually impaired will be able to retrieve information about the obstacle in an efficient manner, allowing them to avoid it. The obstacle size in pixels^2 in the image is determined by the "Bounding Box" property of the predefined function "Regionprops" of MATLAB that locates and delimits obstacles with an encompassing box to retrieve their dimensions. The calibration curve established, giving the proportionality between the area in pixels^2 of the obstacle in the image and its real surface area in m^2, made it possible to evaluate the actual dimensions.

2 Materials and Methods

The stereoscopic image pairs are captured by means of an acquisition device that consists essentially of two digital cameras mounted on a linear and rigid support, which allows to properly stabilize and quickly adjust the distance between both objectives through a horizontal rail. The algorithm provides information on the depth and size of the obstacle in a reasonable time, enabling the blind or visually impaired to make the necessary decisions for his movement before approaching the obstacle. The proposed algorithm has been coded and tested in a MATLAB environment, Firstly, on nearby obstacles and small dimensions (Cameras-obstacle distance ranging from 1 m to 4 m with a step of 5×10^{-1} m); Front surface area varies between $4,200 \times 10^{-4}$ m^2 and $4,148 \times 10^{-2}$ m^2. Thereafter, the test is carried out on distant obstacles; Cameras-obstacle distance ranging from 5 m to 30 m with a step of 5 m. In this case, the area of the treated obstacles varies from $1,1088$ m^2 to $1,836 \times 10^1$ m^2.

2.1 Automatic Obstacle Detection

The algorithm used introduces a new obstacle extraction technique that requires three steps: The input image is first segmented into homogeneous color elements by "super pixel" exploiting the SLIC method. Next, we compute the contrast levels that evaluate the uniqueness and spatial distribution of these elements. Finally, we derive a binary mask that distinctly separates the obstacle of interest from the image background.

2.1.1 Image Segmentation Using SLIC

The pixels provide extremely local information that can not be used without intensive processing. By contrast, superpixels produce less localized information that contributes to the partition of an image into significant regions. The SLIC method adapts a k-means clustering approach [14] to efficiently generate compact and uniform superpixels that have a similar color distribution, subsequently forming an over-segmentation of an image. It performs a grouping of pixels in $5D$ space defined by the values L, a, b of the color space "L^*, a^*, b^*", which comprises all perceptible colors by the human eye and the pixel coordinates x, y. "L^*, a^*, b^*" space is used to capture the colors so that two colors perceived as similar are adjacent; very different colors are distinctly distant.

In Fig. 1, we present the main steps of the SLIC algorithm implemented in *MATLAB*:

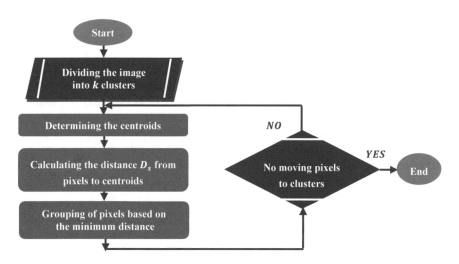

Fig. 1. Schema of the *SLIC* algorithm using $k - means$ clustering

1. Dividing the image of n pixels into k groups sharing the same color, called clusters, with an approximate size of each cluster is n/k pixels. Thereafter, for superpixels of equal size, there is a cluster center at each grid interval $S = \sqrt{n/k}$.

2. Taking initial values for the cluster centroids $C_i = [L_i, a_i, b_i, x_i, y_i]^T$ where i varies from 1 to k.
3. Determining the distance of each pixel to the centroids using Euclidean distance D_s defined as follows [14]:

$$D_s = d_{lab} + \frac{m}{S} d_{xy}$$

With:

$$d_{lab} = \sqrt{(l_i - l_j)^2 + (a_i - a_j)^2 + (b_i - b_j)^2} \quad and \quad d_{xy} = \sqrt{(x_i - x_j)^2 + (y_i - y_j)^2}$$

The variable m is introduced to control the compactness of a superpixel. The cluster is compact and spatial proximity is accentuated for a high value of m.

4. Grouping of pixels according to the minimum distance (search for the closest centroid).
5. Identifying new cluster centroids based on new accessions and repeating steps 3 and 4 until the pixels are no longer moved to another cluster. Thus, the computation of $k - means$ clustering reaches its stability and no iteration is required.

2.1.2   Local Contrast Difference

To extract the object of interest, the superpixels of the regions created by *SLIC* process are treated using the difference in average intensity between a pixel and its neighbors, each element is therefore represented by the average contrast of pixels belonging to it. Thus, depending on the uniqueness in the spatial distribution of superpixels, we produce a contrast map that uniformly covers the objects of interest and systematically separates the front and the background.

Objects of all sizes are present in the segmented image. To take the great foreground object constituting the obstacle, we label the connected regions, then we extract the surface of each. Through a thresholding according to these surfaces, we find the largest region by eliminating all the others. To determine the surface $(L \times l)$ in *pixels²* that translates the white pixels number belonging to the object, the binary image is first scanned pixel by pixel in the four directions from the barycenter position of the object in search of the transition between its shape and the background, then the difference between the coordinates of transition pixels at horizontal and vertical level is calculated in order to count the pixels number in the width "L" as well as in the length "l".

2.2   Depth Calculation for an Obstacle

The depth calculation of a 3D point from stereoscopic images is almost immediate if these images are rectified, as can be seen in the figure below (Fig. 2):

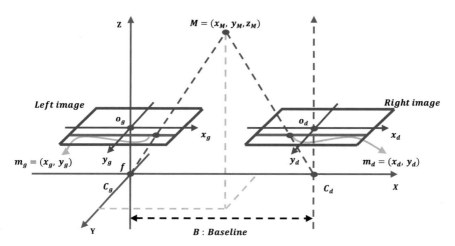

Fig. 2. Depth calculation of a 3D point for an aligned configuration of the cameras

A 3D point M has the projections:

$m_g = (x_g, y_g)$ in the frame $(o_g x_g y_g)$ associated with the left image,
$m_d = (x_d, y_d)$ in the frame $(o_d x_d y_d)$ associated with the right image.

The horizontal position x_g (resp. x_d) of the two points m and m', is given by:

$$x_g = f\frac{x_M}{z_M} \quad \text{and} \quad x_d = f\frac{x_M - B}{z_M}$$

So we have:

$$x_g - x_d = \left(f\frac{x_M}{z_M}\right) - \left(f\frac{x_M - B}{z_M}\right)$$

The difference between positions of the two projections in both images represents the disparity that can be found as shown in the following equation:

$$x_g - x_d = f\frac{B}{z_M}$$

The depth Z of 3D point is inversely proportional to the disparity measured on the image, defined as $d = x_g - x_d$.

$$z_M = \frac{B \times f}{x_g - x_d} = \frac{B \times f}{d} \quad \text{With:} \quad x_g - x_d > 0$$

2.3   Stereoscopic Matching

Pixel matching into binocular stereovision consists to locate in two images of the same scene, taken at different positions, the pixel pairs which correspond to the projection of a same element.

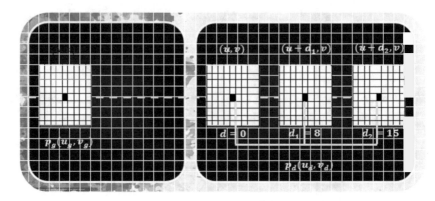

Fig. 3. Principle of the block matching algorithm

We can estimate the disparity by intensity correspondence of the two stereoscopic images. However, since there are often many intensity repetitions in one or more zones, pixel matching is unreliable. To solve this problem, the most commonly used method is to add neighborhood information to increase robustness. It is indeed the Block Matching algorithm, that we have employed in this work and which can be approached as an optimization problem, aimed at finding the best block within a research area. The corresponding costs for each pixel determine the probability of a correct match, the pixel selected is the one with the lowest cost (see Fig. 3).

A similarity measure is the degree of correlation between two windows, the correlation window centered on the pixel $p_g(u_g, v_g)$ to match from the left image and the sliding window centered on a potential pixel $p_d(u_d, v_d)$ located on the right epipolar of the other image I_d, of identical sizes $(2N+1) \times (2N+1)$. In this procedure, the disparity is obtained by minimizing the **SAD** criterion.

2.3.1 Disparity Map

To visualize the result of matching, we use an image called a disparity map. Each pixel of the map represents disparity amplitude: An important disparity will make the pixel appear lighter, and vice versa. Figure 4 shows an example of a stereoscopic image pair and the disparity map associated.

2.3.2 Refinement Process

In matching if the resolution is low, an error of a pixel generates an error of several meters in the distance calculation. Disparities values returned by the Block Matching are all integers, so that there is no smooth transition between regions of different disparities. In contrast the subpixel technique has been introduced to improve the disparity estimation quality for processed pixel. For example, if we have a 256×256 data matrix that we display at a resolution of 64×64 pixels, the data available to construct the image statistically exceed one sample per pixel. This means that instead of getting a location in the image in terms of pixel coordinates (x, y), which implies integer values for x and y, the location is calculated to eventually give positions of fractional pixels. Consequently, the positions of the correspondents are no longer integers causing a large amount of incorrect calculations, but real numbers [15].

Fig. 4. Stereoscopic image pair with the associated disparity map

The problem of discerning optimal disparity estimates and eliminating false matches is solved by the dynamic programming algorithm which distinguishes the best match between two sets of pixels on homologous epipolar lines [16].

In view of the speed problems, the pairs of acquired stereoscopic images are sampled by means of a multi-resolution pyramid structure which is faster in the stereoscopic matching than the other without multi-resolution because the search range in each level of the pyramidal structure is considerably reduced [17].

2.4   Determination of the Obstacle Dimensions

The path of a blind or visually impaired person is always encumbered by obstacles of different forms which must be detected and identified in order to eliminate the risks of shocks. For this, we actively interested in estimating the actual dimensions of the surrounding obstacle whatever the distance, which is possible if we know its size in the image.

We use the *MATLAB* function Regionprops on the segmented image to locate obstacles with the Bounding Box property, defined as the smallest rectangle that can contain the dimensions or frame of an obstacle. The selection box around the obstacle with all abducted holes is drawn from its barycenter (gravity center) identified as follows: in a first step, we calculate the sum of 1 in the binary image which constitutes the total number of white pixels inside the obstacle form. In the next step, we return for all non-zero pixels the sum of its rows and columns coordinates in two variables, and consequently we determine the x and y barycenter coordinates while dividing the sum found by the integral number of 1.

Once the barycenter is identified, we take the column passing through this barycenter and we scan from its pixels each image line alternately to right and left, then as soon as the pixel value becomes 0, we store the coordinates of the last white pixel. Thus, we search in both directions for the pixel coordinates belonging to the obstacle contour giving a maximum distance with its gravity center. The same procedure is repeated for the line that also passes through the barycenter, scanning in this case is done upwards and downwards. The coordinates of the four sides are used to draw the rectangle surrounding the obstacle in order to derive its width and height.

3   Results and Discussion

To facilitate the treatment in the Z-depth computation process of a $3D$ point, we must have rectified image pairs as inputs, which makes it possible to optimize and considerably simplify the pixels matching phase of the stereoscopic images. Epipolar rectification is a geometric transformation applied to the two stereoscopic images, and consists in switching to a sensor configuration that matching pixels lie on the same line (Fig. 5).

Fig. 5. Epipolar rectification: the first line shows images couple and its rectified version is on the second line

The red and blue lines in the above example show that the same point in the real world is on the same horizontal line in each rectified image. The rectification of stereoscopic images pair, therefore, enables to restrict the search space of the matches in a very important manner: we pass from an initial two-dimensional search space $2D$ to a 1D monodimensional space, because the disparity is only a difference of columns.

3.1   Evaluation of the Stereoscopic Algorithm

We tested in *MATLAB* a series of images for stereoscopic pair in order to determine the threshold distance below which the object distance between the images pair is apparent and easily detectable. The results obtained show that the shift observed in the stereoscopic image is all the smaller as the objects are located very far from the sensor. Thus, the distinction between the images becomes imperceptible for 26 m. As a result, we can consider the distance 25 m as depth limit from which the objects have no visible disparity with the naked eye. While, the presented algorithm could discriminate the positional difference for each element of a scene up to a distance of 30 m.

3.2   Comment on the Main Results

Detection threshold corresponding to the smallest distance between the obstacle and the stereoscopic sensor for which the algorithm is able to determine the position and size of the obstacle is 0,5 m. When the distance falls below this value, both cameras do not cover exactly the same parts of the scene.

We can easily detect an obstacle located at a distance less than 0,5 m by bringing the optical axes of two cameras to converge on this obstacle, but in this case it is necessary to apply a transformation at both images so that their epipolar lines are parallel and aligned. In our algorithm, we avoided this costly solution in computing time by adopting a parallel configuration of cameras pair.

Detection limit giving the greatest distance from the obstacle to be detected is set at 30 m. Beyond this maximum detection value, we have not noticed a tangible difference between stereoscopic images pair, which means a disparity equal to 0 making it impossible to evaluate the camera-obstacle distance.

A valid comparison of the performance and efficiency of our algorithm with those of stereoscopic algorithms adopted by existing navigation systems [18–20] is difficult due to a lack of detailed information on their accuracy and processing time of the stereoscopic image pair, but we can estimate that the proposed method in this paper has the advantage over previous ones to determine in addition to the obstacle distance, the dimensions of the latter. It offers a large detection range, which provides time for the visually impaired to act. The treatment does not concern the entire image but only the region of interest constituting the obstacle, an approach that significantly helps reduce the execution time of the algorithm.

3.3   Detection Capacity and Robustness of the Presented Algorithm

Spontaneous walking speed of the visually impaired is approximately equal to 0,6 m/s. Furthermore, when walking the safety distance to be checked to avoid obstacles was defined as 2,5 m.

Our algorithm puts 2,89 s on a Windows 7 machine in 64 bits for an Intel Core $i7$ processor of 3, 40 GHz with 8192 MB of *RAM*. This means that during this time the visually impaired traverses a distance of 1, 734 m, at which is added the distance 0,6 m traveled during the reaction time (approximately 1 s), but which however remains less than 2,5 m. Accordingly, the developed algorithm might be able to guide mobility of visually impaired with adequate security without jerky behavior.

4 Conclusions and Perspectives

The performance of an obstacle detection system is inherent to the power of the stereoscopic vision algorithm. In this paper, we presented a new algorithm for the time-limited and optimal evaluation of the obstacle position and its size, based on Block Matching method. The capacity of the proposed method was verified by tests in a number of real stereoscopic sequences captured from a stereoscopic camera under different scenarios and with various changes in obstacle scales. Thus, we can estimate that our navigation aid algorithm for people who are blind or partially sighted has been developed, while taking into account the time constraint, whereby compliance is as important as result accuracy.

At the current stage in the advancement of our study, different perspectives can be offered. The first point to improve concerns the experiments on the algorithm execution time, and is considered as an unavoidable step in the battle of producing distance data between the stereoscopic camera and obstacles, in real time. Another interesting research track to explore is to test our solution in real conditions. Thus, we are actively considering converting information on potential obstacles obtained by image processing methods to sound signals or vibrations.

References

1. Doush, I.A., Alshatnawi, S., Al-Tamimi, A.-K., Alhasan, B., Hamasha, S.: ISAB: integrated indoor navigation system for the blind. Interact. Comput. **29**(2), 181–202 (2017)
2. Tsirmpas, C., Rompas, A., Fokou, O., Koutsouris, D.: An indoor navigation system for visually impaired and elderly people based on radio frequency identification (RFID). Inf. Sci. **320**(1), 288–305 (2015)
3. Dhod, R., Singh, G., Singh, G., Kaur, M.: Low cost GPS and GSM based navigational aid for visually impaired people. Wirel. Pers. Commun. **92**(4), 1575–1589 (2017)
4. Khenkar, S., Alsulaiman, H., Ismail, S., Fairaq, A., Jarraya, S.K., Ben-Abdallah, H.: ENVISION: assisted navigation of visually impaired smartphone users. Procedia Comput. Sci. **100**, 128–135 (2016)
5. Song, J., Song, W., Cheng, Y., Cao, X.: The design of a guide device with multi-function to aid travel for blind person. Int. J. Smart Home **10**(4), 77–86 (2016)
6. Bahadir, S.K., Koncar, V., Kalaoglu, F.: Smart shirt for obstacle avoidance for visually impaired persons. In: Smart Textiles and their Applications. A volume in Woodhead Publishing Series in Textiles, pp. 33–70 (2016)
7. Rizzo, J.-R., Pan, Y., Hudson, T., Wong, E.K., Fang, Y.: Sensor fusion for ecologically valid obstacle identification: building a comprehensive assistive technology platform for the visually impaired. In: 2017 7th International Conference on Modeling, Simulation, and Applied Optimization, ICMSAO 2017, 26 May 2017, Article Number 7934891 (2017)
8. Rodríguez, A., Javier Yebes, J., Alcantarilla, P.F., Bergasa, L.M., Almazán, J., Cela, A.: Assisting the visually impaired: obstacle detection and warning system by acoustic feedback. Sensors (Switzerland) **12**(12), 17476–17496 (2012)
9. Al-Mutib, K., Mattar, E., Alsulaiman, M.: Implementation of fuzzy decision based mobile robot navigation using stereo vision. Procedia Comput. Sci. **62**, 143–150 (2015). Proceedings of the 2015, International Conference on Soft Computing and Software Engineering (SCSE 2015)

10. Yang, Y., Gao, M., Zhang, J., Zha, Z., Wang, Z.: Depth map super-resolution using stereo-vision-assisted model. Neurocomputing, Part C **149**, 1396–1406 (2015)
11. Elachhab, A., Mikou, M.: Feasibility study of a navigation aid system for the visually impaired, determination of the depth and size of an obstacle by stereoscopic images processing. Int. J. Eng. Res. Technol. (IJERT) **3**(4), 180–185 (2014)
12. Elachhab, A., Mikou, M.: Obstacle detection algorithm based on stereoscopic images, a navigation aid system for the visually impaired. Int. J. Sci. Eng. Res. **6**(4), 872–878 (2015)
13. Elachhab, A., Mikou, M.: Obstacle detection based on stereoscopic vision and the local contrast difference, autonomous navigation assistance for visually impaired. J. Next Gener. Inf. Technol. (JNIT) **7**(3), 1–12 (2016)
14. Achanta, R., Shaji, A., Smith, K., Lucchi, A., Fua, P., Süsstrunk, S.: SLIC Superpixels, EPFL Technical, report 149300, June 2010
15. Arce, E., Marroquin, J.L.: High-precision stereo disparity estimation using HMMF models. Image Vis. Comput. **25**(5), 623–636 (2007)
16. Hu, T., Qi, B., Wu, T., Xu, X., He, H.: Stereo matching using weighted dynamic programming on a single-direction four-connected tree. Comput. Vis. Image Underst. **116**, 908–921 (2012)
17. Aribi, W., Kalfallah, A., Elkadri, N., Farhat, L., Siala, W., Daoud, J., Bouhelel, M.S.: Évaluation de Techniques Pyramidales de Fusion Multimodale (IRM/TEP) d'Images Cérébrales. In: 5th International Conference: Sciences of Electronic, Technologies of Information and Telecommunications, Tunisia, 22–26 March 2009
18. Kammouna, S., Parseihianc, G., Gutierreza, O., Brilhaulta, A., Serpaa, A., Raynala, M., Oriolaa, B., Macea, M.J.-M., Auvrayc, M., Denisc, M., Thorpeb, S.J., Truilleta, P., Katzc, B. F.G., Jouffraisa, C.: Navigation and space perception assistance for the visually impaired: the NAVIG project. IRBM **33**(2), 182–189 (2012). Numéro spécial ANR TECSANTechnologie pour la santé et l'autonomie
19. Costa, P., Fernandes, H., Martins, P., Barroso, J., Hadjileontiadis, L.J.: Obstacle detection using stereo imaging to assist the navigation of visually impaired people. Procedia Comput. Sci. **14**, 83–93 (2012). Proceedings of the 4th International Conference on Software Development for Enhancing Accessibility and Fighting Info-exclusion (DSAI 2012)
20. Bujacz, M.: Sonified stereovision travel aid for local navigation, the project is financed by the National Centre of Research and Development of Poland in years 2010–2013 under the grant NR02-0083-10. http://www.naviton.pl/

A Simultaneous Topic and Sentiment Classification of Tweets

Doaa Hassan[(⊠)]

Department of Computers and Systems, National Telecommunication Institute,
5 Mahmoud El Miligy Street, 6th district-Nasr City, Cairo, Egypt
doaa@nti.sci.eg

Abstract. Social networks have been an emerging technology for communication among billions of users. One of the most popular social networks is Twitter. The popularity of Twitter comes from its simplicity
since it allows users to exchange messages of short length that does not
exceed 140 characters and takes the form of tweets. In this paper, we propose a model for performing a classification of tweets posted by the Twitter user based on a mixture of the topic and sentiment of those tweets.
The proposed approach is new in that it creates a model that combines
the processes of topic and sentiment classification of tweets simultaneously. Therefore, with this model, one can categorize tweets according to
their topics and simultaneously assign them into different sentiments categories. The topic of the tweets in the basic experiment of the proposed
approach is classified into five main different categories including: "political", "commercials", "educational", "religious", and "sportive". Meanwhile, the sentiment of those tweets is classified into three main different
categories including "positive", "negative", "neutral". The effectiveness
of the proposed approach is demonstrated on a real dataset that consists of various extracted tweets with different categories of topics and
opinions. The empirical results show that our approach is very powerful
in categorizing tweets according to topics and simultaneously assigning
them into different sentiments categories.

Keywords: Text mining · Topic classification · Sentiment analysis
Twitter

1 Introduction

Social networks have been an emerging technology for communication among
billions of users. One of the most popular social networks is Twitter. The popularity of Twitter comes from its simplicity as it allows users to exchange messages of short length that does not exceed 140 characters and take the form of
tweets that lets users express everything. The Twitter API is used for accessing
all data specified by communication on Twitter. The nature of communication
among users provides a very powerful way for collecting a large sets of data. An
example of such communication is users that are following the user who posts

© Springer International Publishing AG, part of Springer Nature 2018
A. Abraham et al. (Eds.): SoCPaR 2017, AISC 737, pp. 24–33, 2018.
https://doi.org/10.1007/978-3-319-76357-6_3

the tweet are allowed to see his/her post. Moreover, various actions are allowed to be performed on the posted tweet including replying the tweet by notifying the user who posted the tweet with the reply, and retweeting a tweet by posting the tweet as if it is a new tweet posted by the user. This can be shown to his/her followers, while keeping the information which the user originally posted. The collected data by communication on Twitter is considered as powerful source of information for detecting users interests, and their opinions regarding various topics discussed on Twitter.

Topic modeling and sentiment analysis are two common natural language processing techniques that have been widely used for processing the large and continuous streams of Twitter data. The main goal of performing topic modeling on Twitter data is to process large collections of Twitter data to detect the main subjectivity of Twitter messages and identify the kinds of topics that are mainly talked about through communication on Twitter. Meanwhile, performing sentiment analysis on Twitter data aims to process the large collection of opinions texts on Twitter in order to detect the sentiment polarity of tweets.

In this paper, we propose an approach for conducting a composite classification of tweets to reveal the major topic of each tweet and associate each topic with a corresponding sentiment polarity. The proposed approach is new in that it creates a model that combines the process of topic and sentiment classification of tweets simultaneously. Therefore, with this model, one can categorize tweets according to topics and simultaneously assign them into different sentiments categories. This approach would be more useful for analyzing sentiments of tweets at the level of subtopics, as a tweet often covers a mixture of subtopics and may hold different opinions for different subtopics. The effectiveness of the proposed approach is demonstrated on a real dataset that consists of various extracted tweets with different categories of topics and opinions. The topics of the tweets posted by Twitter users in the basic experiment of the proposed approach is classified into five main different categories including: "political", "commercials", "educational", "religious", and "sportive". Meanwhile, the sentiment of those tweets is classified into three main different categories including "positive", "negative", and "neutral".

The reminder of this paper is organized as follows. In Sect. 2, we describe the related work. In Sect. 3, a general description of the main proposed approach of this paper is presented. In Sect. 4, the basic experiment of this paper is conducted and the performance of the proposed approach is evaluated. Finally, in Sect. 5, we conclude the paper with some directions for future work.

2   Related Work

Recently, there has been a research direction for classifying Twitter microblogs messages into topical categories using text mining techniques for topic modeling [1,2]. Another research direction has been focusing on performing the sentiment analysis of tweets via classification [3]. Our proposed work in this paper aims to simultaneously integrate both directions by combining the classification of tweets topics and their sentiment analysis in one process. In the following, we summarize the recent research work that addressed the combination of topic classification and sentiment analysis of tweets.

Batista and Ribeiro [8] presented an approach in which the topic classification and sentiment analysis were performed over Spanish Twitter data. The process was performed in the context of a satellite event of the SEPLN 2012 conference that consists of tweets written in Spanish. Each tweet is automatically labeled in terms of sentiment polarity and the corresponding topic using the polarity lexicons for sentiment analysis and the Latent Dirichelet Allocation for topic detection [12]. However, the processes of sentiment and topic classification of each tweet were performed separately and no combination in one process was presented.

In [9], a research work was presented for evaluating several preprocessing, feature extraction methods (including n-grams, stemmers and lemmatizers, types of word, valence shifters, negations, search engines, link processing, and Twitter hashtags) for the purpose of classifying Spanish tweets according to sentiment and topic using different classifiers. However, similar to the work presented in [8], no simultaneous combination of topic and sentiment classification was presented.

In [10], the authors performed topic classification of tweets texts through considering the context of sentiment tweets when the performing sentiment analysis of the collected tweets. Each tweet in the training dataset has been manually labeled with a label representing the content of the tweet and a label representing the sentiment of the tweet. The determination of tweet topics as well as their sentiment analysis was performed using the multinominal Naive Bayes classifier. However this work also integrates topic classification and sentiment analysis in two successive classification stages.

Our presented work in this paper is different from the previous research works mentioned above as it integrates the topic classification and sentiment analysis simultaneously. This has been inspired by the work of Mei et el. [15], in which they proposed a probabilistic model for capturing the mixture of topics and sentiments simultaneously. However their model was not empirically applied and tested on Twitter Microblogs.

3   Basic Approach

The proposed approach in this paper aims to build a model for a simultaneous classification of topics and sentiments of tweets posted by Twitter users. The model consists of two phases as shown in Fig. 1. In the first phase, a batch of tweets is labeled with composite label that actually combines two labels: one for presenting the topic, while the other one presents the sentiment of that topic. The methodology that have been used for determining the topic and sentiment labels of the tweet corpus will be described in details in the next section. In second phase, the labeled tweets from the first phase are preprocessed by removing special characters such as #, ?, @, etc. and removing urls, extra white space, and stop words (e.g., the, I, on, in, etc.). Next the preprocessed corpus of tweets is used as an input dataset for training and testing a supervised classifier. The classifier would be able to detect which topic each tweet is about, as well as detecting whether the tweet has presented this topic neutrally, positively, or negatively.

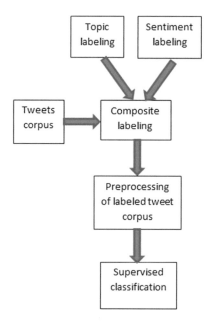

Fig. 1. A model for a simultaneous classification of topics and sentiments of tweets.

4   Basic Experiment

4.1   Dataset Collection

The dataset that has been used in the basic experiment of this paper consists of smaller sets of tweets with various topics from different categories including "political", "commercials", "educational", "religious", and "sportive". Each set of tweets (i.e., tweet corpus) have been collected by extracting the tweets posted by a user and containing a top hashtag that is commonly used for searching about a certain topic. Therefore, the main topic of hashtag presents the topic category of this tweets set. For example, we have collected tweets with top hashtags related to sportive topics that consist of a word or a phrase preceded with a hash symbol #, (e.g. #cycling, #Wimbledon, etc.). In other words, the top hashtags that refer to a particular topic are used as a way for categorizing tweets topics. This approach was previously presented in [5].

The collection program has been executed several times. Each time, the top hashtag used for fetching tweets posted by the user is changed, resulting in a dataset containing 10 sets of tweets of five different topic categories and collected from different time periods. This dataset contains 2000 tweets.

Due to the Twitter terms of service agreement [4] that prevents the distribution of a copy of tweets, the dataset has not been made available for public. The main reason for this is to allow the user who posted the tweet to omit it, which can't be achieved if making copies of the tweets is allowed.

Each tweet in the extracted datasets has been manually tagged with a composite tag that consists of two labels separated by "-", the first one describes the main topic of the tweet, while the second one describes the sentiment of the tweet. An example of some extracted tweets labeled with a composite label is shown in Table 1.

Table 1. An example of tweets labeled with composite labels.

Tweet instance	Topic-sentiment label
#Donald Trump's #first state visit to China is successful, visiting Vice-Premier Liu Yand https://t.co/JcNdLt3jV5	political-positive
#ICYMI Robert Mugabe says #Donald Trump has a mental illness and speaks before he thinks. https://t.co/CfSBRTdW16	political-negative
Any #Wimbledon based companies want to get involved in a local event? Check out @WeAreBobblehat #prrequest #brandswanted #R	sportive-neutral

As for the topic label, it is assigned to each tweet in the tweet corpus to match the topic of the top hashtag used as a search term for gathering the tweets in this corpus. As for the sentiment label, it is assigned according to the approach presented in [14]. This approach determines the sentiment of each word in the tweet by matching it with two manually crafted lexicons of positive and negative words with predefined polarity scores. Then the final polarity score for each tweet is calculated by summing the polarity scores of all words contained in that tweet.

The dataset has been used for training and testing the experimental classifier in 10-fold cross validation process [20]. In this process 10% of samples of the dataset is randomly selected for testing the classification accuracy against the obtained accuracy when training the classifier with the remaining 90% of the samples of the dataset.

4.2 Experimental Settings

We have used R [17] and its integrated development environment RStudio [18] for extracting tweets that contains a topic specific hashtag from Twitter, where the searchTwitter function in R has been used to fetch tweets with a topic specific hashtag from Twitter. Also R has been used for preprocessing of tweets texts before classification as Fig. 1 shows and it will be described in the next section. Weka [19], a free data mining software tool has been used for topic-sentiment classification of tweets using various generated supervised classifiers.

4.3   Tweets Text Preprocessing

The first step in tweets text preprocessing is to clean the extracted dataset of tweets by removing url, special characters (e.g., #, @, ?, etc.) and stopword words (e.g., the generic redundant terms/words in all extracted tweets such as: you, the, is, etc.). Next, a corpus of a pre-processed tweets (a set of tweet documents) is created and stored in a dataframe. This dataframe is written to CSV file, that is a suitable input to Weka for performing classification of tweets.

Second, the Weka StringToWordVector filter [21] is used to convert the string attribute that represents each tweet document in the CSV file into a set of attributes representing word occurrence/frequency in that document. This filter has various options that allow to do the following:

- use Term Frequency-Inverse Document Frequency (TF-IDF) weighting scheme [13] to determine the value of each attribute by counting how many times each word in each tweet document is presented.
- apply stemming mechanism for reducing words into their root form by removing their prefixes, suffixes and infixes.
- determine the minimum frequency of words/terms in the term matrix[1].

4.4   Classification

The goal of the classification stage is to assign one class that is composed of two sub-classes to each tweet. The first subclass describes the main topic of the tweet, while the other one describes the sentiment of the tweet.

For performing tweets classification, we have used Weka FilteredClassifier [21] that allows to run any arbitrary supervised classifier on the tweet corpus that has been passed through the StringToWordVector filter mentioned in the previous subsection.

Two classifiers are constructed to perform classification of the tweets, including Naive Bays, and Support Vector Machine (SVM) classifiers. Both classifiers have been chosen because they were commonly used by other approaches presented in Sect. 2 for topic and sentiment classification of tweets. All unigrams and bigrams (i.e. all words and word pairs respectively) in the input text were considered as features for training the constructed classifiers [6].

The classifiers are trained and tested using the dataset of tweets that has been manually tagged with the composite label as introduced early in this section. As a result, each tweet is then classified by the generated classifiers to determine the composite label that expresses the topic of the tweet as well as its sentiment.

4.5   Performance Evaluation

The overall accuracy has been used as a measure for evaluating the classification performance as well as the precision, recall and F-measure. Table 2 shows the accuracy of experimental classifiers when using either unigrams or bigrams as features for training the classifiers.

---

[1] In this term matrix, a row refers to a candidate word/term and a column refers to a tweet document.

Table 2. The accuracy of experimental classifiers when using either unigrams or bigrams as features for training the classifiers.

N-gram	1	2
Classifier	Accuracy	
SVM	84.85%	56.20%
Naïve-Bayes	69.45%	50.45%

Clearly, the results in Table 2 show that SVM outperforms Naïve-bayes classifier either when using unigrams or bigrams as features for training the constructed classifier. Moreover, the results show that using unigrams as a feature for training the classifier improves the accuracy, while using bigrams as a feature achieves lower accuracy for both constructed classifiers. However, using bigrams creates a better understanding of context when compared with only using unigrams for sentiment analysis. The negation is considered as an example to show this property. For example, the bigram "not successful" is an indication of negative sentiment, while the unigrams "not" and "successful" may incorrectly indicate a positive sentiment. Therefore, and due to space constraints, we only present the precision, recall and F-measure of all composite classes for each of both construed classifiers in case of using bi-grams as features for training the classifier. Tables 3, and 4 show such presentations for SVM, and Naïve Bayes classifiers respectively.

Table 3. Precision, recall and F-measure of SVM classifier.

Precision	Recall	F-Measure	Class
0.99	0.616	0.759	political-neutral
0.886	0.544	0.674	political-positive
1	0.357	0.526	political-negative
0.889	0.293	0.44	commercial-positive
0.932	0.356	0.515	commercial-neutral
0	0	0	commercial-negative
0.9	0.26	0.403	educational-positive
0.251	0.996	0.401	educational-neutral
0	0	0	educational-negative
1	0.877	0.935	religious-negative
0.961	0.601	0.739	religious-neutral
0	0	0	religious-positive
0.981	0.599	0.744	sportive-positive
0.871	0.279	0.422	sportive-neutral
0	0	0	sportive-negative
0.827	0.562	0.592	Weighted Avg.

Table 4. Precision, recall and F-measure of Naïve Bayes classifier.

Precision	Recall	F-Measure	Class
1	0.502	0.668	political-neutral
0.574	0.544	0.559	political-positive
0.833	0.357	0.5	political-negative
0.941	0.195	0.323	commercial-positive
0.259	0.922	0.404	commercial-neutral
0	0	0	commercial-negative
0.844	0.26	0.397	educational-positive
0.967	0.208	0.343	educational-neutral
0.143	0.077	0.1	educational-negative
0.994	0.855	0.919	religious-negative
0.99	0.493	0.658	religious-neutral
0	0	0	religious-positive
1	0.503	0.669	sportive-positive
0.406	0.365	0.385	sportive-neutral
0	0	0	sportive-negative
0.768	0.504	0.524	Weighted Avg.

It should be noticed that it is difficult to compare the performance of our approach with the recent approaches investigated in the Related Work section such as [9,10]. This is because our approach performs a simultaneous classification of topic and sentiment of tweets in one stage, while those approaches performed topic and sentiment classification in two separated stages. However, we have used the classification accuracy as a basis for comparing our approach to those approaches by comparing the accuracy of our approach to the average accuracy of topic and sentiment classification of those approaches. Table 5 shows the results of such comparison, when using bi-gram as feature for training the classifier. As the results in Table 5 show, our approach outperforms other approaches either for SVM or Naïve Bayes classifiers.

Table 5. A comparison between the accuracy of our approach and those presented in [8–10]

Approaches	Our approach	Approach in [9]	Approach in [10]
Classifier	Accuracy		
SVM	56.20%	47.40%	–
Naïve Bayes	50.45%	44.24%	31.70%

5   Conclusions and Future Work

In this paper we have proposed a new model for a simultaneous classification of tweets based on their categorical topics and their sentiment polarities (specified by the different opinions regarding each topic category). This is achieved by simultaneously combining topic and sentiment classification of tweets that users post on Twitter. In the first stage of the proposed model, a simultaneous topic and sentiment classification is performed by labelling the dataset that consists of various collected tweet corpus with a composite label. Such a label consists of two parts. The first part refers to the label of the context/topic which is manually decided by deciding the topic of each tweet corpus based on the topic of the top hashtags queried for searching about that corpus. The second part of the label refers to the label of tweet sentiment which is decided by matching each tweet in the corpus to two lexicons of positive and negative words then calculating the final sentiment score. Therefore, the dataset is labeled with a composite label that describes both of the topic of each tweet and the opinion about it. In the second stage, this dataset is used to train and test various classifiers from different categories in order to enable them to detect the topic as well as the sentiment of tested tweets simultaneously. To demonstrate the effectiveness of our approach, we have tested it on a real dataset that consists of various extracted tweets with different categories of topics and sentiment, then reported the classification performance. The empirical results have shown that our approach is very powerful for categorizing tweets according to their topics and simultaneously assigning them into different sentiments categories.

The future work may take several directions. First, we need to test the degree of relevance of the extracted tweet data to the original search parameter (i.e., top hashtag) on the accurate classification of topic and hence the accuracy of simultaneous topic-sentiment of tweets. Second, we are looking forward to testing the performance of our approach when the topic classification for training the classifier is performed using one of the common topic modeling techniques [2, 16]. Finally, we are looking forward to extracting the topic life cycle with modeling its dynamics and its corresponding sentiments in order to deeply understand how the opinions about a specific topic change over time.

References

1. Yang, S., Kolcz, A., Schlaikjer, A., Gupta, P.: Large-scale high-precision topic modeling on Twitter. In: Proceedings of KDD14, 24–27 August 2014, New York, NY, USA (2014)
2. Alghamdi, R., Alfalqi, K.: A survey of topic modeling in text mining. Int. J. Adv. Comput. Sci. Appl. (IJACSA) **6**(1) (2015)
3. Kharde, V.A., Sonawane, S.S.: Sentiment analysis of Twitter data: a survey of techniques. Int. J. Comput. Appl. (IJCA) **139**(11), 5–15 (2016)
4. Twitter Terms of Service. https://twitter.com/en/tos

5. Llewellyn, C., Grover, C., Alex, B., Oberlander, J., Tobin, R.: Extracting a topic specific dataset from a Twitter archive. In: Proceedings of the 19th International Conference on Theory and Practice of Digital Libraries (TPDL 2015), pp. 364–367 (2015)
6. Balahur, A.: Sentiment analysis in social media texts. In: Proceedings of the 4th Workshop on Computational Approaches to Subjectivity, Sentiment and Social Media Analysis, Atlanta, Georgia, June 2013, pp. 120–128. ©2013 Association for Computational Linguistics (2013)
7. Wang, S., Manning, C.: Baselines and bigrams: simple, good sentiment and topic classification. In: Proceedings of ACL (2012)
8. Batista, F., Ribeiro, R.: Sentiment analysis and topic classification: case study over spanish tweets. In: Proceedings of TASS 2012, Satellite Event of the SEPLN 2012 Conference, 7 September, Valencia, Spain (2012)
9. Anta, A.F., Chiroque, L., Morere, P., Santos, A.: Sentiment analysis and topic detection of Spanish tweets: a comparative study of NLP techniques. J. Proces. del Leng. Nat. **50**, 45–52 (2013)
10. David, J.: Sentiment and topic classification of messages on Twitter and using the results to interact with Twitter users. Examensarbete 30 hp, Uppsala University, Mars 2016
11. Bak, J.Y., Lin, C.-Y. Oh, A.: Self-disclosure topic model for Twitter conversations. In: Proceedings of the Joint Workshop on Social Dynamics and Personal Attributes in Social Media, Baltimore, Maryland, USA, 27 June 2014, pp. 42–49. ©2014 Association for Computational Linguistics (2014)
12. Blei, D.M., Ng, A.Y., Jordan, M.I.: Latent Dirichlet allocation. J. Mach. Learn. Res. **3**, 993–1022 (2003)
13. Manning, C.D., Raghavan, P., Schütze, H.: Introduction to Information Retrieval. Cambridge University Press, Cambridge (2008)
14. Liu, B., Hu, M., Cheng, J.: Opinion observer: analyzing and comparing opinions on the web. In: Proceedings of the 14th International World Wide Web conference (WWW-2005), 10–14 May 2005, Chiba, Japan (2005)
15. Mei, Q., Ling, X., Wondra, M., Su, H., Zhai, C.: Topic sentiment mixture: modeling facets and opinions in weblogs. In: Proceedings of the 16th International Conference on World Wide Web (WWW 2007), 8–12 May 2007, Banff, Alberta, Canada, pp. 171–180 (2007)
16. Ashique Mahmood, A.S.M.: Literature Survey on Topic Modeling. Dept. of CIS, University of Delaware (2013)
17. The R Project for Statistical Computing. https://www.r-project.org/
18. RStudio open source and enterprise-ready professional software for R. https://www.rstudio.com/
19. Hall, M., Frank, E., Holmes, G., Pfahringer, B., Reutemann, P., Witten, I.H.: The WEKA data mining software: an update. J. SIGKDD Explor. Newsl. **11**(1), 10–18 (2009)
20. Tan, P.-N., Steinbach, M., Kumar, V.: Introduction to Data Mining, 1st edn. Addison-Wesley, Boston (2005)
21. Hidalgo, J.M.G.: Text Mining in WEKA: Chaining Filters and Classifiers, January 2013

Local Directional Multi Radius Binary Pattern

Novel Descriptor for Face Recognition Application

Mohamed Kas[1(✉)], Youssef El Merabet[1], Yassine Ruichek[2],
and Rochdi Messoussi[1]

[1] Laboratoire LASTID, Département de Physique, Faculté des Sciences,
Université Ibn Tofail, Kénitra, Morocco
mohamed.kas@uit.ac.ma,
y.el-merabet@univ-ibntofail.ac.ma,
messoussi@gmail.com
[2] Le2i, UMR 6306, CNRS Université de Bourgogne Franche-Comté, UTBM,
90010 Belfort, France
yassine.ruichek@utbm.fr

Abstract. Face recognition becomes an important task performed routinely in our daily lives. This application is encouraged by the wide availability of powerful and low-cost desktop and embedded computing systems, while the need comes from the integration in too much real world systems including biometric authentication, surveillance, human-computer interaction, and multimedia management. This article proposes a new variant of LBP descriptor referred as Local Directional Multi Radius Binary Pattern (LDMRBP) as a robust and effective face descriptor. The proposed LDMRBP operator is built using new neighborhood topology and new pattern encoding scheme. The adopted face recognition system consists of three stages: (1) face detection and alignment to normalize the input images to a common form if needed; (2) feature extraction using the proposed descriptor in order to calculate the histogram, which represents the feature vector and (3) face recognition through a supervised image classification task using the simple K-Nearest Neighbors classifier. Simulated experiments on ORL, YALE and FERET under different illumination or facial expression conditions indicate that the proposed method outperforms other texture descriptors and other existing works of the literature.

Keywords: Face recognition · Feature extraction · Face descriptor
LBP · Classification · LDMRBP

1 Introduction

The face of a human being carries enough descriptions useful to identity a person and its emotional state. Over the course of its development, the human brain acquires highly specialized areas dedicated to the analysis of the facial images. Similar to the human learning, the researchers had developed intelligent systems based on the machine learning that was born from pattern recognition and the theory that computers can perform specific tasks just by learning and without being programmed. These tasks

© Springer International Publishing AG, part of Springer Nature 2018
A. Abraham et al. (Eds.): SoCPaR 2017, AISC 737, pp. 34–48, 2018.
https://doi.org/10.1007/978-3-319-76357-6_4

include face recognition, which is an interesting and challenging problem and impacts important applications in many areas as network security, computer vision and content indexing and retrieval.

Face recognition has recently increased thanks to the machine learning and artificial intelligence. However, it is still not accurate all the time especially when the environment is variable. The ability to recognize and to classify facial images correctly depends on a variety of variables such as lighting conditions, pose, facial expressions, image quality and the structure of the recognition system including preprocessing steps, feature extraction, and the classification algorithm.

The typical configuration of face recognition systems includes three main and necessary stages: *Stage #1:* face detection and alignment to crop and normalize face images. This stage may include also preprocessing techniques for illumination problem that attempt to normalize all face images to a common illumination situation. *Stage #2:* face description; this stage takes the normalized face image as an input and by applying a specific description algorithm, it outputs the corresponding feature vector who contains the image description, which is the key to a successful face recognition system if the representation is more accurate. *Stage#3:* face classification, here we perform a matching of a query image with those images in the learning set, this classification is based on the feature vectors.

In applications that require automated system with higher performance and matching accuracy like face recognition, the face descriptor is a distinguishing element and a key issue. If the used descriptor in the feature extraction phase cannot effectively represent faces, even the sophisticated classifier may fail to recognize and classify the query face image. A good descriptor is expected to reduce the intra-person variance while exploiting the variance between different persons to easily classify and recognize the probe image. During several decades of research and development in face recognition, numerous face representation and description methods have been proposed [1–4]. These approaches can be generally categorized into two classes: holistic and local methods.

The holistic face representation approaches are based on exploiting subspaces learning, which usually send the gray-scale face images as inputs to a classifier. There are many methods for this class In the literature, exist two representative subspace learning Principal component analysis (PCA) [5] and linear discriminant analysis (LDA) [36]. Unlike the holistic methods, the class of local face representation attempts mainly to assemble and construct the information locally within a face image to describe the image itself. LBP-based facial image classification is considered as one of the most popular and successful applications at the present time. The LBP operator was proposed by Ojala et al. [7] in 2002 as a texture descriptor, moreover the local binary pattern (LBP) texture method has provided significant results in many applications due to its better tolerance against illumination changes in real world applications including face recognition. Another equally important property is its lower computational simplicity and the length of its feature vector, which makes it able to analyze images in challenging real-time conditions. This success of the LBP in many applications gave birth of a vast number of LBP variants that have been proposed and continue to be proposed. Tan and Triggs introduced in [8] a three-values operator called local ternary patterns (LTP) for face recognition. The basic idea is to encode the difference between the center pixel and its neighbor pixels by three values (1, 0 or −1) giving a user specified threshold. Inspired by LBP, higher order local derivative patterns (LDP) were proposed in [9], with applications in face recognition. Although LDP operator provides more detailed information, it is more sensitive to noise than LBP descriptor. Recently, Yang et al. [28], inspired by Weber's Law, proposed adaptive local ternary patterns (ALTP) feature descriptor based on an automatic strategy selecting the threshold for LTP. Srinivasa and Chandra [6] proposed a dimensionality reduced local directional pattern (DR-LDP) which computes single code for each block by X-ORing the LDP codes obtained in a single block.

Motivated by these works, this paper proposes a novel local binary descriptor referred as Local Directional Multi Radius Binary Pattern (LDMRBP). The proposed operator computes and describes the relationship between the referenced pixel and its neighbors on a 5 × 5 pixel block size by encoding gray-level difference based on two-level radius $(R = 1$ and $R = 2)$ and multi direction angles: 0°, 45°, 90° and 135°. The idea behind is to combine the radius with the angle to extract more detailed and discriminating description. To make the thresholding process more accurate, each pixel is compared to a specific mean of its 3 × 3 neighbor pixels within the original 5 × 5 neighborhood. Our descriptor is characterized by its effectiveness in face recognition application using the simplest K-Nearest Neighbor classifier configured with City Block distance measure. This effective performance is proven on three challenging and widely used databases of the literature such as ORL database, FERET database with various poses and angles and the Yale database who trials the classification process with massive changes in the illuminations, lighting conditions and the variable background.

The main contributions of this paper can be summarized and briefed in the following points:

- We propose a novel variant of the LBP texture descriptor referred as Local Directional Multi Radius Binary Pattern (LDMRBP), which combines two topological dimensions to describe a given pixel keeping a low computational requirements and simple conception.
- The performance of the proposed LDMRBP operator and its performance stability are evaluated on three benchmark databases. To make this investigation more meaningful, we record the average classification rate over 10 random splits and according to different numbers of training images.
- The performance of the proposed LDMRBP descriptor is compared to several state-of-the-art LBP variants and systems.

The rest of this paper is arranged as follows. Section 2 presents briefly some texture descriptors as related works. Section 3 presents the proposed LDMRBP operator in addition to the face recognition framework. Comprehensive experimental results and comparative evaluation are given in Sect. 4. Section 5 concludes the paper and proposes some future research directions.

2   Related Works

In recent years, LBP has made a significant progress in various application including face recognition, texture classification and image retrieval. This progress motivated researchers to develop more variants of LBP operator. In this section, we present a quick review of the basic LBP descriptor and some methods with different neighborhood topology and sampling.

2.1   The Basic Local Binary Patterns Operator

Ojala et al. [7] proposed LBP descriptor as a powerful feature for texture classification. It was later used for solving face recognition due to its ability to represent face image. It's defined as a gray scale invariant texture measure, derived from a general definition of texture in a local neighborhood.

The original LBP operator labels the pixels of an image with decimal numbers referred to as LBP codes that encode and describe the local structure around each pixel. It proceeds thus, as illustrated in Fig. 1: Each pixel is compared with its eight neighbors in a 3 × 3 neighborhood by thresholding the center pixel value; The resulting strictly negative values are encoded with 0 and the others with 1; A binary number is obtained by concatenating all these calculated binary codes in a clockwise direction starting from the top-left one and its corresponding decimal value is used for labeling.

In Fig. 2, the neighborhood was expanded to capture dominant feature with large-scale structures. The neighborhood can be denoted by a pair *(P, R)* where *P* is the sampling points on a circle of radius of *R*. Therefore, there are 2^P different output values.

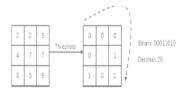

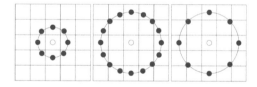

Fig. 1. 3 × 3 computed bloc

Fig. 2. The circular (8, 1), (16, 2) and (8, 2) neighborhoods

If the coordinates of the center pixel are (x_c, y_c) then the coordinates of his P neighbors (x_p, y_p) on the edge of the circle with radius R can be calculated as follows:

$$x_p = x_c + R \cdot \cos\left(\frac{2\pi p}{P}\right) \tag{1}$$

$$y_p = y_c + R \cdot \sin\left(\frac{2\pi p}{P}\right) \tag{2}$$

the final code of the gray-scale image I is calculated considering the sign $S(z)$ of the differences between central pixel value $I(x_0)$ and its P neighbors pixel values $I(x_k)$ sampled on a circle of given radius R, the kernel function is:

$$LBP_P^R(x_0) = \sum_{k=1}^{P} S\{I(x_k) - I(x_0)\} \cdot 2^{k-1} \tag{3}$$

Where

$$S(z) = \begin{cases} 1, z \geq 0 \\ 0, z < 0 \end{cases} \tag{4}$$

As the LBP values assigned to the image pixels for P = 8 are in the range [0, 255], they can be visualized in form of 8-bit images. Figure 3 depicts exemplary image (a), corresponding maps of LBP values (b) and the calculated histogram (c).

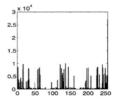

Fig. 3. (a) Input image, (b) LBP image, (c) LBP histogram

2.2 Directional Local Binary Patterns and Local Binary Patterns by Neighborhoods

Recently, Kaya et al. [10] proposed two LBP variants, referred to as *Directional local binary patterns* (dLBPα) and *Local binary patterns by neighborhoods* (nLBPd). dLBPα operator is based on determining the neighbors in the same orientation over central pixel angle parameter, which may take 0°, 45°, 90° or 135° as illustrated in Fig. 4. nLBPd operator is based on defining the neighbors used in the comparison which can be done not only with sequential neighbors but also inside the neighbors defined by a distance parameter d. Figure 5 illustrates an example of nLBPd calculation with d = 1 and d = 2. The kernel function is given by the following equation:

$$nLBP_d(x_0) = \sum_{k=1}^{P} f\{I(x_k), I(x_{(k+d)mod(P)})\} \cdot 2^{k-1} \tag{5}$$

Where

$$f(P_i, P_j) = f(x) = \begin{cases} 1, P_i > P_j \\ 0, P_i \leq P_j \end{cases} \tag{6}$$

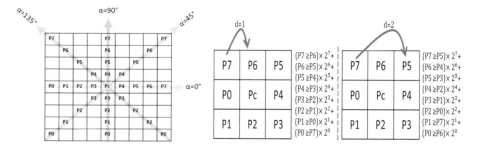

Fig. 4. Displacement vectors of dLBPα

Fig. 5. The overview of relations between neighbors in the nLBPd

3 Local Directional Multi Radius Binary Pattern

In this section we present a discriminative, highly robust and yet efficient method for facial images feature extraction and description, namely Local Directional Multi Radius Binary Pattern.

The conception of the proposed LDMRBP descriptor is based on two essential aspects: neighborhood topology and pattern encoding.

3.1 Neighborhood Topology

We adopted a 5×5 block size in LDMRBP descriptor manipulating 16 pixels. In order to capture the maximum information in the facial image, these pixel samples cover the, at same time, four orientations which are $[0°, 45°, 90°, 135°]$ and two radiuses $[R = 1, R = 2]$ as shown in Fig. 6. This propriety allows to describe the variance intra-person in according to the changes that can be found in first row neighbors and the ones of the second row.

As can be seen from Fig. 6, we define two rows of pixels; pixels of A row where the distance between them and the central pixel Pc is $R = 1$ and those who are far by $R = 2$ belong to the B row. These 16 pixel samples will be arranged into two sampling groups with 8 pixels in each one. This division is based on combining the pixels of one radius along two directions from level A and the second radius with the other directions from level B as illustrated in Fig. 7.

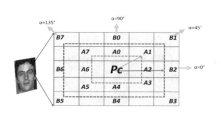

Fig. 6. Local sampling topology of LDMRBP

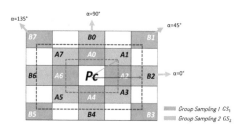

Fig. 7. The two groups of sampling adopted in LDMRBP

We can define it also in a second way, the GS_1 contains the pair pixels of A and the odd ones of B while the GS_2 includes the odd pixels of A and the pair samples of B. This process is described in the following equations (cf. Eqs. (7) and (8)):

$$GS_1 = \{A_{2i}, B_{2i+1}\} \tag{7}$$

$$GS_2 = \{A_{2i+1}, B_{2i}\} \tag{8}$$

The rest of the pixels within the 5×5 neighborhood are used in order to calculate the mean of each pixel from A row. Based on the above neighborhood topology, eight mean values are obtained and labelled $\{M_{A0}, \ldots \ldots M_{A7}\}$. These mean values will be used in the pattern encoding phase.

3.2 Pattern Encoding

After defining the neighborhood topology, which permits to define pixels that will be used in the modeling, we present, as a second step, the pattern encoding scheme of the proposed LDMRBP descriptor. The local information is encoded using two encoders named Ec_1 and Ec_2 based, in addition to the central pixel, on the two groups of samples and the eight mean values $\{M_{A0}, \ldots \ldots M_{A7}\}$ defined previously. The two encoders adopt the basic thresholding function where each pixel, except the central pixel, is compared to the mean which it belongs to. Note that, the central pixel is compared to the mean M_{Ec_1} of the pair of pixels of A row and M_{Ec_2} mean in the case of the odd pixels $GS_1 GS_2$ will be associated to Ec_1 and Ec_2 respectively where they are given as:

$$Ec_1 = f(P_c, M_{Ec_1}) \times 2^8 + \sum_{p=4}^{7} f(A_{2i}, M_{A_{2i}}) \times 2^p + \sum_{p=0}^{3} f(B_{2i+1}, M_{A_{2i+1}}) \times 2^p \tag{9}$$

$$Ec_2 = f(P_c, M_{Ec_1}) \times 2^8 + \sum_{p=4}^{7} f(A_{2i+1}, M_{A_{2i+1}}) \times 2^p + \sum_{p=0}^{3} f(B_{2i}, M_{A_{2i}}) \times 2^p \tag{10}$$

Where

$$f(x,y) = \begin{cases} 1, & x \geq y \\ 0, & otherwise \end{cases} \tag{11}$$

$$M_{Ec_1} = \frac{\sum_{i=0}^{3} A_{2i}}{4} \tag{12}$$

$$M_{Ec_2} = \frac{\sum_{i=0}^{3} A_{2i+1}}{4} \tag{13}$$

The final LDMRBP code obtained for each pixel of the image, is the concatenation of the two features generated by the two cross encoders Ec_1 and Ec_2.

$$LDMRBP_{Pc} = \langle Ec_1, Ec_2 \rangle \tag{14}$$

3.3   Implementation

Similar to most state-of-the-art face recognition methods, the proposed system, as shown in Fig. 8, involves several steps.

First, the images of each dataset are preliminarily divided into 10 random splits generated at each number of training images; each split contains two sub-sets, one for the training and the other for the test. Secondly, the code images are obtained using the proposed LDMRBP operator. After that, the obtained code images are further divided into N non-overlapping blocks and a histogram of patterns is generated for each block. The histogram bins of all blocks are concatenated to form the final descriptor of the whole images. Finally, the images of the test set are classified through the nearest-neighbor rule (1-NN) with City Block distance given in Eq. (15). The procedure is repeated 10 times, each time with new subdivision into training and validation sets, and the accuracy obtained with each subdivision is recorded.

$$CityBlock(x, y) = \sum_{j=1}^{k} |x_j - y_j| \tag{15}$$

4   Experimental Analysis and Discussions

In order to evaluate the performance and the effectiveness of the proposed LDMRBP descriptor and to validate its performance stability, we conducted a comparative analysis through the discussed framework above using three challenging databases. In addition to our descriptor, this section investigates also the performance of 20 LBP

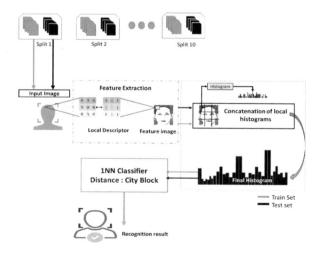

Fig. 8. Overview of the face recognition system

Table 1. Evaluated LBP variants

Abbreviation	Complete name	Year	Ref
BGC1	Binary Gradient Contours (1)	2011	[11]
BGC2	Binary Gradient Contours (2)	2011	[11]
BGC3	Binary Gradient Contours (3)	2011	[11]
CLBP_M	Completed LBP (Magnitude)	2010	[12]
DBC	Directional Binary Code	2010	[13]
DCLBP	Multi-Scale Densely Sampled Complete LBP	2012	[14]
dLBPα	Directional Local Binary Patterns	2015	[10]
DRLBP	Dominant Rotated Local Binary Patterns	2016	[15]
IBGC1	Improved binary gradient contours (1)	2013	[16]
ILTP	Improved Local Ternary Patterns	2010	[17]
LBP	Local Binary Patterns	2002	[7]
nLBPd	Local binary patterns by neighborhoods	2015	[10]
LDN	Local Directional Number Pattern	2013	[18]
LDP	Local Directional Pattern	2013	[9]
LQCH	Local Quantization Code Histogram	2016	[19]
LTP	Local Ternary Patterns	2007	[8]
MMEPOP	Mag Maximum Edge Position Octal Pattern	2015	[20]
OC_LTP	Orthogonal Combination of LTP	2014	[21]
QBP	Quad Binary Patterns	2016	[22]
SMEPOP	Sign Maximum Edge Position Octal Pattern	2015	[20]
WLD	Weber Local Descriptor	2010	[23]

variants to make the comparison more meaningful. The evaluated LBP-like methods are listed in Table 1.

4.1 Datasets

ORL: AT&T Laboratories Cambridge database was founded in 1986 better known as ORL [24]. This database has 400 images, including 40 individuals and 10 images per person. The size of each image is 92×112.

FERET: The FERET face image database [25] is a result of the FERET program, which was managed by the US Department of Defense through the DARPA Program. The database contains 1564 sets of images giving a total of 14126 images that includes 1199 individuals. We use a subset that contains 1400 images of 200 persons (7 images per person) as adopted in [26].

YALE: Yale Face Database contains 165 grayscale images of 15 individuals. There are 11 images per subject, one per different facial expression or configuration. These images are used to analyze the performance of the texture models under noisy conditions, different pose and illumination conditions.

4.2 Experimental Results and Discussions

The Experimental Analysis

Experiment #1: This experiment is conducted on FERET database. We setup a subset containing not only the frontal images but also samples with various orientations to test the robustness of the proposed method against this challenge. In this database, the histogram computing and concatenation are done over 9 blocks for all the tested methods. Table 2 summarizes the obtained average face recognition accuracies recorded over the ten splits. It can be clearly seen that the LDMRBP outperformed all the evaluated LBP-like methods of the literature. Indeed, it achieves a higher recognition rate of 99.65% over the ten splits, vs 98.09%. with DCLBP descriptor, vs 98.04% with BGC1, etc. Note that, a common evolution in the average accuracy for all the description methods is noticed when we enlarge the train set with more samples. Having an average accuracy above 90% in a challenging database like Feret is a satisfactory result, where the proposed descriptor started to achieve higher accuracies (above 92.5%) after adopting 3 images in the train set. From Table 5, we find that the accuracy recorded by the proposed surpassed those reported in the literature even if in our case we consider ten random splits and not one.

Table 2. Average accuracy (%) recorded over FERET

Descriptor	Train\Test					
	1\6	2\5	3\4	4\3	5\2	6\1
BGC1	56.37	76.84	86.41	92.35	95.4	97.34
BGC2	54.5	74.9	85.21	91.19	94.97	97.39
BGC3	53.7	74.19	84.35	90.67	94.6	96.83
CLBP_M	51.39	71.17	82.37	89.38	93.34	95.93
DBC	47.77	68.46	80.09	87.44	92.81	94.77
DCLBP	56.92	77.62	88	93.84	95.98	98.09
DLBPa	56.55	75.31	85.19	91.19	93.97	95.78
DRLBP	54.48	71.69	81.63	86.67	91.43	93.42
IBGC1	57.71	78.21	88.03	93.53	96.76	98.04
ILTP	55.78	76.09	86.67	93.32	95.58	97.79
LBPd	54.56	74.66	85.44	91.64	95	97.34
LBP	54.36	74.85	85.49	91.42	94.97	97.04
LDN	44.5	64.54	76.22	84.1	88.79	93.17
LDP	45.58	65.66	77.75	85.16	90.23	93.32
LQCH	43.87	62.45	73.22	81.54	86.11	89.7
LTP	54.86	74.53	85.09	91.74	95.03	96.98
MMEPOP	48.19	69.22	81.14	88.46	92.91	96.08
OC_LTP	49.54	68.78	80.05	86.8	90.85	93.87
QBP	50.18	68.95	79.71	86.75	90.38	93.17
SMEPOP	52.01	72.77	84.08	90.45	94.17	96.88
WLD	62.54	78.84	87.05	92.35	94.95	97.09
Proposed	61.76	82.72	92.55	96.78	98.57	99.65

Experiment #2: This experiment is conducted on Yale database; In this experiment, the objective is to demonstrate the performance and the effectiveness of the proposed LDMRBP operator against the huge change of the illumination conditions and also the uncontrolled environment between the subjects. In this database, the feature extraction is performed using 100 blocks and we calculate the histogram on each one. Table 3 summarizes the classification performance of the proposed descriptor and those of the state-of-the-art methods. It is apparent from Table 3 that the LDMRBP descriptor prove a great success in Yale database. Indeed, it achieves for the train\test set configuration, accuracies above 90% and it's the only method to reach 100% over the random splits considering 7 images in the train and 4 as a probe samples. Moreover, we remark that the proposed method presents the requested stability according to the train \test configurations while the rest of the evaluated LBP variants are affected by these configurations. For example, the performance of LBP operator decreases by 3% at 8\3 configuration comparing to the accuracy obtained in 7\4.

Experiment #3: The final experiment is curried out on ORL database. This database is widely used in the literature to test new methods and approaches due to its equilibrated number of subjects and the samples per each one, in addition to the uncontrolled

Table 3. Average accuracy (%) recorded over ORL

Descriptor	Train\Test								
	1\9	2\8	3\7	4\6	5\5	6\4	7\3	8\2	9\1
BGC1	77.94	87.62	95.32	95.33	98.3	98.5	99.25	99.5	99.5
BGC2	75.44	86.38	94	94.96	97.75	97.81	98.83	99.25	99.5
BGC3	74.36	85.81	93.82	94.63	97.5	98.25	99.08	99.38	99.5
CLBP_M	72.83	85.25	91.71	93.29	96.5	97	98.58	97.88	99.5
DBC	70.81	81.97	91.04	92.96	97.1	96.88	98.08	98.62	99.75
DCLBP	78.58	88.53	94.54	96.5	98.2	98.44	99.58	99.5	100
dLBPa	72.19	84.44	90.07	92.83	95.25	96.81	98.08	98.38	99.25
DRLBP	63.56	78.16	87.11	88.71	93.25	95	95	96.88	98
IBGC1	79.78	88.66	96	96.08	98.55	98.69	99.33	99.75	100
ILTP	78.58	89.41	95.29	96.58	98.7	98.25	99.25	99.62	100
LBPd	75.64	86.19	94.14	94.79	97.5	97.44	98.75	99.12	99.25
LBP	76.75	87.12	94.71	95.37	98	98	98.67	99.38	99.25
LDN	64.33	76.31	86.36	89.46	91.8	93.44	95.58	96.88	97
LDP	66.03	79.25	88.61	90.79	94.6	95.5	98.08	98.38	98.25
LQCH	57.89	71.03	82.18	84.25	89.75	91.19	93.42	95.12	94.75
LTP	77.83	88.59	95.11	95.83	98.55	98.44	99.42	100	100
MMEPOP	68.5	81.56	90.14	91.92	94.9	95.69	97.67	98.12	99.25
OC_LTP	71.69	83.44	91.04	93.38	96.65	97.88	98.5	98.75	99.75
QBP	72.06	84.16	91.79	93.17	95.9	96.88	98	99	99.5
SMEPOP	74.78	85.72	93.93	94.79	96.95	98.19	98.58	99	99.5
WLD	79.61	88.91	94.86	96	97.55	97.75	98.58	99.25	99.25
Proposed	87.03	95.78	99.04	99.67	100	100	100	100	100

Table 4. Average accuracy (%) recorded over YALE

Descriptor	Train\Test									
	1\10	2\9	3\8	4\7	5\6	6\5	7\4	8\3	9\2	10\1
BGC1	75.13	86.96	88.92	88	89.67	90.13	91.83	91.78	89.33	94
BGC2	75.27	88.22	90.67	88.95	90.67	91.2	93.67	94	90.67	96
BGC3	76.2	88.74	90.42	89.33	90.67	90.8	92.5	92.67	90	94.67
CLBP_M	74.93	87.19	91.25	91.52	92	93.33	93.5	91.56	93.33	92
DBC	76.87	87.41	89.92	88.38	90.22	90.53	93	92.44	91.33	94
DCLBP	80.6	91.63	94.58	94.57	94.44	94.8	95.33	95.56	96	98
dLBPa	80.67	90.96	94.67	94.57	96.11	95.73	98.33	98.89	98	100
DRLBP	72.93	84.67	88.67	88.29	88.56	89.07	93	89.78	90	91.33
IBGC1	76.13	87.26	90	89.14	91	90.8	92.33	92.89	89.33	94
ILTP	77.6	88.74	89.75	90	91.89	93.33	93.5	92.22	91.33	92.67
LBPd	75.27	87.78	90.42	88.57	91	90.67	92.33	92	89.33	94.67
LBP	78.4	90.59	92.75	91.81	93.22	92.93	95.33	95.78	92.67	97.33
LDN	75.73	86.96	90.75	90.67	92.13	94.17	93.78	91.67	94.23	96.67
LDP	74.13	86	89.17	89.05	89.44	90.93	92.17	89.56	88.67	90
LQCH	39.4	50.3	54.17	57.52	58.44	56.93	60	58.67	60.67	62
LTP	77.67	88.3	89.42	89.24	90.78	92.67	93	92.22	92.33	92
MMEPOP	75.4	87.48	92.25	92.29	94.33	94.13	96.5	96.22	94	97.33
OC_LTP	72.33	84.96	88.58	86.48	86	86.67	87.83	86.67	85.67	86
QBP	73.47	86.15	88.75	88.48	89	89.33	91.5	88.67	89.33	88.67
SMEPOP	74.93	87.26	89.08	88.29	89	90.8	92	92.22	89.33	94
WLD	59.73	73.93	78.67	78.48	79.22	80.93	80.5	83.56	79.67	87.33
Proposed	91.53	96.59	97.58	98.29	99	98.93	100	100	100	100

environment and the frontal image. We divide each image into 9 blocks, then we compute the histogram over each one. The challenge herein is to achieve 100% average accuracy with a number of reference images as minimum as possible. Table 4 reports the average accuracy of each evaluated method. It emerges from this Table that the LDMRBP operator reaches the highest accuracies against the tested state-of-the-art methods. Furthermore, it achieves 99.04% at 3\7 config outperforming the rest by at least 5% and the most important thing that can be noticed is the 100% average accuracy recorded over the half\half (Train\Test) configuration and over the random splits. The LTP reaches also an encouraging results in ORL database where it realizes 100% average accuracy at 8\2 configuration while three other variants record this accuracy until the 9\1 setup.

Comparaison with state-of-the-art works

In this subsection, we compare the best performance of the proposed LDMRBP method recorded in Section of experimental analysis with existing state-of-the-art systems. Table 5 summarizes comparison results with several existing works. The following findings can be drawn from the analysis of this Table:

Table 5. Comparison with state-of-the-art systems

Database	Reference	Reported accuracy (Train\Test)
Feret	[27]	65.4% (2\5)
	[26]	90.7% (6\1)
	[28]	92.25% (5\2)
	[29]	95.40% (6\4 & 100 subjects)
	[30]	91.2% (3\1 & 250 subjects)
	Proposed	**99.65% (6\1)**
Yale	[31]	45.33% (1\10)
	[32]	88.56% (8\3)
	[33]	75.56% (8\3)
	[34]	75% (-)
	[35]	86% (6\11)
	Proposed	**100% (7\4)**
Orl	[27]	89.5 (5\5)
	[6]	97.62% (5\5)
	[31]	71.94% (1\9)
	[32]	98.52% (8\2)
	[26]	96% (5\5)
	[28]	100% (8\2)
	[29]	98.9% (6\4)
	Proposed	**100% (5\5)**

- FERET: It easily found that the accuracy recorded by the proposed LDMRBP operator surpassed those reported in the literature even if in our case we consider ten random splits and not one. Indeed, the proposed method allowed, as shown in Table 5, to achieve, 99.65% correct classification rate, Vs 95.40% in [29], Vs 91.2% in [30], etc.
- YALE: As can be noticed from Table 5 and according to our best knowledge, we are the only to achieve 100% average accuracy with a configuration around 63.6% for the train and 36.4% as a probe set.
- ORL: According to state-of-the-art accuracies given in Table 5, [28] achieved also 100% correct classification rate but only over one split and with 8 images in the train not only 5 in our case.

5   Conclusion

In this paper, we proposed a novel local descriptor named LDMRBP for face recognition. The proposed LDMRBP operator is based on combining two different concepts of the neighborhood sampling which are the direction and the radius. A comprehensive evaluation of the proposed LDMRBP descriptor is performed on three challenging representative widely-used face datasets, with comparison to 20 LBP-like methods and several state-of-the-art systems. Jugged by the analysis of the experimental results, the proposed LDMRBP model is always the top ranked descriptor and outperforms all the evaluated LBP-like methods and state-of-the-art systems. In future works, we plan to further explore the discriminative power of our method in other challenging computer vision applications such as gender classification and texture classification.

References

1. Cai, D., He, X., Han, J., Zhang, H.: Orthogonal laplacianfaces for face recognition. IEEE Trans. Image Process. **15**, 3608–3614 (2006)
2. Wang, X., Li, Z., Tao, D.: Subspaces indexing model on grassmann manifold for image search. IEEE Trans. Image Process. **20**, 2627–2635 (2011)
3. Tzimiropoulos, G., Zafeiriou, S., Pantic, M.: Subspace learning from image gradient orientations. IEEE Trans. Pattern Anal. Mach. Intell. **34**, 2454–2466 (2012)
4. Lu, J., Liong, V., Zhou, X., Zhou, J.: Learning compact binary face descriptor for face recognition. IEEE Trans. Pattern Anal. Mach. Intell. **37**, 2041–2056 (2015)
5. Pentland, M.: Eigenfaces for recognition. J. Cogn. Neurosci. **3**(1), 71–86 (1993)
6. Srinivasa, P.R., Chandra, M.P.V.S.S.R.: Dimensionality reduced local directional pattern (DR-LDP) for face recognition. Expert Syst. Appl. **63**, 66–73 (2016)
7. Ojala, T., Pietikainen, M., Maenpaa, T.: Multiresolution gray-scale and rotation invariant texture classification with local binary patterns. IEEE Trans. Pattern Anal. Mach. Intell. **24**, 971–987 (2002)
8. Tan, X., Triggs, B.: Enhanced local texture feature sets for face recognition under difficult lighting conditions. IEEE Trans. Image Process. **19**, 1635–1650 (2010)
9. Jabid, T., Kabir, M., Chae, O.: Local directional pattern (LDP) for face recognition. In: 2010 Digest of Technical Papers International Conference on Consumer Electronics (ICCE), pp. 329–330 (2010)
10. Kaya, Y., Ertuğrul, Ö., Tekin, R.: Two novel local binary pattern descriptors for texture analysis. Appl. Soft Comput. **34**, 728–735 (2015)
11. Fernández, A., Álvarez, M., Bianconi, F.: Image classification with binary gradient contours. Opt. Lasers Eng. **49**, 1177–1184 (2011)
12. Guo, Z., Zhang, L., Zhang, D.: A completed modeling of local binary pattern operator for texture classification. IEEE Trans. Image Process. **19**, 1657–1663 (2010)
13. Zhang, B., Zhang, L., Zhang, D., Shen, L.: Directional binary code with application to PolyU near-infrared face database. Pattern Recogn. Lett. **31**, 2337–2344 (2010)
14. Ylioinas, J., Hadid, A., Guo, Y., Pietikäinen, M.: Efficient image appearance description using dense sampling based local binary patterns. In: Asian Conference on Computer Vision, pp. 375–388 (2012)
15. Mehta, R., Egiazarian, K.: Dominant rotated local binary patterns (DRLBP) for texture classification. Pattern Recogn. Lett. **71**, 16–22 (2016)
16. Fernández, A., Álvarez, M., Bianconi, F.: Texture description through histograms of equivalent patterns. J. Math. Imaging Vis. **45**, 76–102 (2013)
17. Nanni, L., Brahnam, S., Lumini, A.: A local approach based on a local binary patterns variant texture descriptor for classifying pain states. Expert Syst. Appl. **37**, 7888–7894 (2010)
18. Rivera, A., Castillo, J., Chae, O.: Local directional number pattern for face analysis: face and expression recognition. IEEE Trans. Image Process. **22**, 1740–1752 (2013)

19. Zhao, Y., Wang, R.-G., Wang, W.-M., Gao, W.: Local quantization code histogram for texture classification. Neurocomputing **207**, 354–364 (2016)
20. Vipparthi, S., Murala, S., Nagar, S., Gonde, A.: Local gabor maximum edge position octal patterns for image retrieval. Neurocomputing **167**, 336–345 (2015)
21. Sun, J., Fan, G., Yu, L., Wu, X.: Concave-convex local binary features for automatic target recognition in infrared imagery. EURASIP J. Image Video Process. **2014**, 1–13 (2014)
22. Zeng, H., Chen, J., Cui, X., Cai, C., Ma, K.-K.: Quad binary pattern and its application in mean-shift tracking. Neurocomputing **217**, 3–10 (2016)
23. Chen, J., Shan, S., He, C., Zhao, G., Pietikainen, M., Chen, X., Gao, W.: WLD: a robust local image descriptor. IEEE Trans. Pattern Anal. Mach. Intell. **32**, 1705–1720 (2010)
24. Samaria, F., Harter, A.: Parameterisation of a stochastic model for human face identification. In: Proceedings of the Second IEEE Workshop on Applications of Computer Vision, 1994, pp. 138–142 (1994)
25. Phillips, P., Moon, H., Rizvi, S., Rauss, P.: The FERET evaluation methodology for face-recognition algorithms. IEEE Trans. Pattern Anal. Mach. Intell. **22**, 1090–1104 (2000)
26. Huang, P., Gao, G., Qian, C., Yang, G., Yang, Z.: Fuzzy linear regression discriminant projection for face recognition. IEEE Access **PP**, 1 (2017)
27. Liu, T., Mi, J.-X., Liu, Y., Li, C.: Robust face recognition via sparse boosting representation. Neurocomputing **214**, 944–957 (2016)
28. Yang, W., Wang, Z., Zhang, B.: Face recognition using adaptive local ternary patterns method. Neurocomputing **213**, 183–190 (2016). Binary Representation Learning in Computer Vision
29. Atta, R., Ghanbari, M.: An efficient face recognition system based on embedded DCT pyramid. IEEE Trans. Consum. Electron. **58**, 1285–1293 (2012)
30. Huang, S., Yang, J.: Linear discriminant regression classification for face recognition. IEEE Signal Process. Lett. **20**, 91–94 (2013)
31. Li, L., Gao, J., Ge, H.: A new face recognition method via semi-discrete decomposition for one sample problem. Optik-Int. J. Light Electron Opt. **127**, 7408–7417 (2016)
32. Huang, S., Zhuang, L.: Exponential discriminant locality preserving projection for face recognition. Neurocomputing **208**, 373–377 (2016)
33. Belahcene, M., Laid, M., Chouchane, A., Ouamane, A., Bourennane, S.: Local descriptors and tensor local preserving projection in face recognition. In: 2016 6th European Workshop on Visual Information Processing (EUVIP), pp. 1–6, October 2016
34. Pan, J., Wang, X., Cheng, Y.: Single-sample face recognition based on LPP feature transfer. IEEE Access **4**, 2873–2884 (2016)
35. Ghinea, G., Kannan, R., Kannaiyan, S.: Gradient-orientation-based PCA subspace for novel face recognition. IEEE Access **2**, 914–920 (2014)
36. Belhumeur, P.N., Hespanha, J.P., Kriegman, D.J.: Eigenfaces vs. fisherfaces: recognition using class specific linear projection. IEEE Trans. Pattern Anal. Mach. Intell. **19**(7), 711–720 (1997)

Content-Based Image Retrieval Approach Using Color and Texture Applied to Two Databases (Coil-100 and Wang)

El Mehdi El Aroussi[✉], Noureddine El Houssif[✉],
and Hassan Silkan[✉]

Laboratory LAMAPI, Department of Mathematics and Computer Science,
Faculty of Science, University Chouaib Doukkali, El Jadida, Morocco
{elaroussi.e, elhoussif.n, silkan.h}@ucd.ac.ma

Abstract. Content-Based Image Retrieval (CBIR) allows to automatically extracting target images according to objective visual contents of the image itself. Representation of visual features and similarity match are important issues in CBIR. Color, texture and shape information have been the primitive image descriptors in content-based image retrieval systems. This paper presents an efficient image indexing and search system based on color and texture features. The color features are represented by combines 2-D histogram and statistical moments and texture features are represented by a gray level co-occurrence matrix (GLCM). To assess and validate our results, many experiments were held in two color spaces HSV and RGB. The descriptor was implemented to two different databases Coil-100 and Wang. The performance is measured in terms of recall and precision; also the obtained performances are compared with several state-of-the-art algorithms and showed that our algorithm is simple, and efficient in terms of results and memory.

Keywords: GLCM · 2-D histogram · Statistical moments · CBIR
HSV · RGB

1 Introduction

The growth of image acquisition devices, storage capacities, in parallel to the lower costs of computer equipment and the availability of high-quality digitization techniques observed in recent years, results in a permanent and considerable production of digital images in different fields, which leads to a constant development of image databases. The database would be irrelevant if information of particular significance and importance could not be found easily. This has prompted a need for data processing development and efficient indexing algorithms to represent and retrieve data in a quick manner. In the development of huge digital media devices, these devices deploy new applications in the area of multimedia information systems, spatial information systems, medical imaging, time-series analysis, image retrieval systems, storage, and compression etc. Indexing and searching for images by content is a promising domain. It offers the possibility for users to access, interrogate and exploit these databases directly using their content. An indexation program is designed as a system liable to

© Springer International Publishing AG, part of Springer Nature 2018
A. Abraham et al. (Eds.): SoCPaR 2017, AISC 737, pp. 49–59, 2018.
https://doi.org/10.1007/978-3-319-76357-6_5

take as an inquiry one image as a reference and the gives back a set of similar images to the reference. This enables the program to sort out images from the very similar to the less similar. The image research system by content seeks to filter the image content automatically thanks to the visual descriptors that represent the multimedia data. The indexation techniques allow organizing and structuring the set of descriptors to operate the search by visual similarities rapidly and efficiently. In this paper, we analyze retrieval performance using color and texture features. The analysis is performed in two color spaces: RGB and HSV. The 2-D histogram and statistical moments are used for color features and gray level co-occurrence matrix (GLCM) features are used for analyzing texture features. It is applied to two databases Coil-100 [20] and Wang [21]. The results of the proposed approach are compared with approach [8, 15, 16] it is shown that the proposed technique provides comparable or better image retrieval results at a much faster rate. The paper is developed as follows: Sect. 2 relayed the literature review pertaining to the recent researches in indexation and Image Content. Section 3 discusses the descriptor under study. Section 4 deals with the results of our experiments; and Sect. 5 draws the conclusions of the proposed approach.

2 Related Work

In this section, a brief review of some important contributions from the existing literature CBIR [1] is presented. In CBIR, image signature plays an imperative role to fabricate an efficient image search. Query image and images found in the repository are qualified as a collection of feature vectors and ranking of the relevant results occur on the basis of common norms, i.e., distance or semantic association by a machine learning technique [2]. Signature development is usually performed through the analysis of texture [3], color [4], or shape [5] or by generating any of these combinations and representing them mathematically [6]. Texture features as powerful visual features are used to capture repetitive patterns of a surface in the images. Color features are extensively used in CBIR. The formation of human identity and recognize objects in the real world is known as an important cue of the shape. Shape features forms have been used for the purpose of retrieving images in many applications [7]. The work of [8] presented a feature extraction approach by generating the curvelet representation of the images. The Curvelet transformations are combined with a vector codebook of region-based sub-band clustering for extraction of dominant color. In this approach, the user-defined image and target images are compared by using the principle of the most similar highest priority (MSHP) and evaluated for the retrieval performance. In the work of [9] a color-texture and dominant color based image retrieval system (CTDCIRS) are proposed and three different features from the images, i.e., motif co-occurrence matrix (MCM), dynamic dominant color (DDC), and the difference between pixels of scan pattern (DBPSP) are offered. Yue et al. [10], they introduced a method for feature extraction which is based on texture-color features. Combination of texture and color features is used for automatic retrieval purpose. Ferreira et al. [11] proposed a secure image encryption algorithm for image retrieval and separately processed the image color information and texture information. This scheme encrypted texture information by the random encryption algorithm and protected color information by using deterministic encryption. In the work of [12], a color histogram based on the wavelet is introduced, which also considers the texture and color component of the images for image retrieval. Xu et al. [13] proposed a triple-bit quantization-based scheme. The scheme assigns a 3-bit to each dimension and applies the asymmetric distance algorithm to re-rank candidates. Although the aforementioned schemes address the privacy issues, the computational burden on users is quite enormous. Liu et al. [14] proposed a scheme based on an encrypted difference histogram (EDH-CBIR) is proposed. Firstly, the image owner calculates the order or disorder difference matrices of RGB components and encrypts them by value replacement and position scrambling. In the work of [15] proposed a new descriptor based on 2-D histogram method with statistical moments by integrating the texture using Gabor filters (GF) and by applying distributed computation to research image. In the work of [16] introduced a mechanism for automatic image retrieval, Therefore, he has applied the bandelet transform for feature extraction, which considers the core objects found in an image. To further enhance the image representation capabilities, color features are also incorporated.

3   Materials and Methods

3.1   Color Feature Extraction

Color property is one of the most widely used visual features in (CBIR) systems. Researchers in this area fall in three steps: (a) definition of adequate color space for a given application, (b) proposal of appropriate extraction algorithms, and (c) study of similarity measures.

3.1.1   Color Histogram
It is the most used descriptor in image retrieval. The color histogram is easy to compute, simple and effective in characterizing the global and the local distribution of colors in an image. The color histogram extraction algorithm uses three steps: partition of the color space into cells, an association of each cell to a histogram bin, and counting of the number of image pixels of each cell and storing this count in the corresponding histogram bin. This descriptor is invariant to rotation and translation the distance between two images can be calculated by

$$D_{Hist}(Q,I) = \frac{\sum_I |Q_{Histo\,i} - I_{Histo\,i}|}{\sum_i Q_{Histo\,i}} \tag{1}$$

3.1.2   Color Moments
Color moments have been successfully used in several retrieval systems. This approach involves calculating the mean, the variance and the third moment for each color channel, for providing a unique number used to index. Color moments have been proved efficient in representing color distributions of images. They are defined as:

$$\text{Mean: } u_i = \frac{1}{N} \sum_{j=1}^{N} p_{ij} \tag{2}$$

$$\text{Standard Deviation: } \sigma_i = \sqrt{\frac{1}{n} \sum_{j=1}^{N} (p_{ij} - u_i)^2} \tag{3}$$

$$\text{Skewness: } S_i = \left(\frac{1}{N} \sum_{j=1}^{n} (p_{ij} - u_i)^3 \right)^{\frac{1}{3}} \tag{4}$$

3.1.3   Combination of Color Descriptor

Since each distance between features is calculated using a different standard, prior to combining these features, we normalize the distances using Iqbal's method [17] this method ensures that the distance is normalized between 0 and 1. After the normalized distances, we use a linear combination of distances.

$$D(Q,I) = W_{Hist} * D_{Hist}(Q,I) + W_{Mom} * D_{Mom}(Q,I) \tag{5}$$

Where W_{Hist2D} and W_{Mom} take a value between 0 and 1. This value can be a new program argument. D2D−M oment is [18] used in [19].

3.2   Texture Feature Extraction

The texture is another important feature of an image that can be extracted for the purpose of image retrieval. Image texture refers to surface patterns that show granular details of an image. The composition of different colors information is also provided in the texture pattern. For example, the different texture pattern can be seen in sky image and block walls image. These two images can be distinguished based on the texture.

3.2.1   Gray Level Co-occurrence Matrix and Two-Order Statistical Parameters

The gray level co-occurrence matrix (GLCM) is a common method for representing the spatial correlation of the pixel grayscale, which mainly describes the image from the interval of adjacent pixels, direction and the extent of variation. The GLCM defines a square matrix whose size represents the probability of the gray value g1 distanced from a fixed spatial location relationship (size and direction) to another gray value g2. Assume that f (i, j) is a 2D gray-scale image, where S is the set of pixels with a certain spatial relation in the region and P refers to the GLCM, which can be expressed as

$$P(i,j) = \frac{\neq \{[(i1,j1),(i2,j2)] \in S[f(i2,j2) = g2]\}}{\neq S} \tag{6}$$

The four types of feature values are employed to extract the texture features of the image:

Angular second moment (ASM)

$$ASM = \sum_i^N \sum_j^N P(i,j)^2 \tag{7}$$

Contrast (CON)

$$CON = \sum_i^N \sum_j^N (i-j)^2 P(i,j) \tag{8}$$

Correlation (COR)

$$COR = -\frac{\sum_i^N \sum_j^N (i-x)(i-y)P(i,j)}{\sigma_x \sigma_y} \tag{9}$$

Entropy (ENT)

$$ENT = -\sum_i^N \sum_j^N P(i,j)LgP(i,j) \tag{10}$$

3.3   Similarity Computation

In the Fig. 1, the way to create the final descriptor is performed in a distributed computing. Indeed, the input image will be indexed as follows: we first create two processes. The first one calculates histogram 2-D with statistical moments, while the second process calculates GLCM texture descriptor. These two processes work in parallel, and finally we combine the results to have the final descriptor.

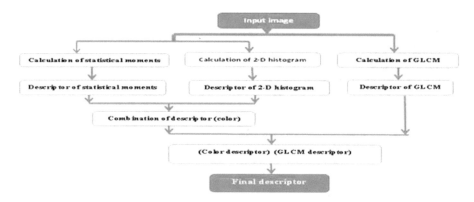

Fig. 1. Architecture of the proposed method.

4 Experimental Results

For the evaluation of our contributions, we used two datasets: The Wang and the COIL-100 (of the size 384 * 256 or 256 * 384 each). We use Wang Database for image retrieval. It is a subset of 1000 images of the Corel stock photo database which have been manually selected and which form 10 classes of 100 images each. The COIL-100 Columbia Object Image Library (COIL-100) is a database of 7200 color images of 100 objects (72 images per object). The objects have a wide variety of complex geometric and reflectance characteristics. The reason for our choice to report the result on these categories is that: these categories are the same semantic groups used by most of the researchers who are working in the domain of CBIR to report the effectiveness of their work [8, 15, 16], so a clear performance comparison is possible in terms of the reported results. Performance evaluation is implemented in Eclipse JEE Mars (JAVA) under a Microsoft Windows 7 environment on an Intel (R) Core (TM) i3 with 2.50 GHz CPU and 4 GB of RAM.

4.1 Image Retrieval

The aim the CBIR is to have relevant images that fit the search request following the need of the user. The more the system provides answers that correspond to the user's need, the more efficient the system. To check the retrieval performance and robustness of the proposed method, several experiments are conducted on COIL-100 and Wang database. We display the best 10 similar images retrieved to prove the performance of our proposed algorithm. The results are presented separately (Figs. 2 and 3). The images found to correspond to the 10 images showing similarities with the query. They are sorted out and posted taking into account distance between the descriptor of the query and the of the images already safeguarded in the database.

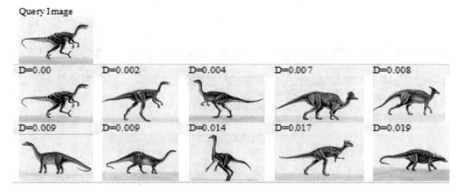

Fig. 2. Example of a query image and similarity search results in the Wang database

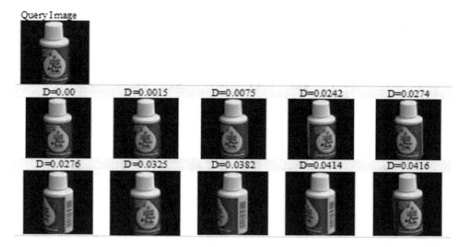

Fig. 3. Example of a query image and similarity search results in the Coil-100 database

4.2   Retrieval Precision/Recall Evaluation

To measure the accuracy of our proposed approach, the standard methods such as Precision and Recall are used. These are the most common measurements used for the evaluation of information and image retrieval systems. Precision or specificity determines the ability of the system to retrieve only those images which are relevant to any query image amongst all of the retrieved images, while Recall rate also known as sensitivity or true positive rate, determines the ability of classifier systems in terms of the model associated with their actual class. For the elaboration of results, top 10 retrieved images against any query image are used to compute the precision and recall. We have reported the average of the results. The proposed method is compared with other state-of-the-art three algorithms in CBIR [8, 15, 16] in terms of precision and Recall. The reason for our choice to compare with these techniques is that these systems have reported their results on the common denomination of the ten semantic categories of Wang and Coil100 dataset. For the Wang database, Table 1 explains the comparison of the proposed system with other comparative systems in terms of precision and Recall in the RGB color space. The best value of mean precision which is 0.823 in the proposed method. This is followed by method [16] with a value is 0.820. The results show that proposed system has performed better than all other systems in terms of average precision obtained. Table 2 describes the performance comparison in terms of precision and Recall rates with the same systems in the HSV color space. From the results, it could be easily observed that the proposed system has the highest recall rates. From the results elaborated in Table 3, the approach [16] provide the best value of mean precision in HSV color space which is 0.955. This is followed by the RGB color space which provides the average precision value 0.947. The performances of mean precision in the approach [8] is 0.949 in HSV color space and value 0.939 in RGB color space. The value of mean precision for the HSV and RGB color spaces, which is 0.917 both for RGB and 0.925 for HSV in the approach [15]. The HSV color

Table 1. Comparison of mean precision/recall obtained by proposed method with other standard retrieval systems in the Wang database and the RGB color space

Method category	Proposed method		[16]		[8]		[15]	
	Precision	Recall	Precision	Recall	Precision	Recall	Precision	Recall
Africa	0.68	0.13	0.65	0.13	0.64	0.13	0.75	0.15
Beach	0.65	0.14	0.7	0.14	0.64	0.13	0.67	0.14
Buildings	0.80	0.14	0.75	0.15	0.70	0.14	0.65	0.13
Buses	0.90	0.19	0.95	0.19	0.92	0.18	0.88	0.18
Dinosaurs	1	0.20	1	0.20	0.99	0.20	1	0.2
Elephants	0.80	0.16	0.80	0.16	0.78	0.16	0.9	0.18
Flowers	0.85	0.20	0.95	0.19	0.95	0.19	0.94	0.19
Horses	0.95	0.18	0.90	0.18	0.95	0.19	1	0.2
Mountains	0.75	0.16	0.75	0.15	0.74	0.15	0.48	0.1
Food	0.85	0.16	0.75	0.15	0.81	0.16	0.6	0.12
Mean	0.823	0.166	0.820	0.164	0.812	0.16	0.79	0.159

Table 2. Comparison of mean precision/recall obtained by proposed method with other standard retrieval systems in the Wang database and the HSV color space

Method category	Proposed method		[16]		[8]		[15]	
	Precision	Recall	Precision	Recall	Precision	Recall	Precision	Recall
Africa	0.7	0.14	0.7	0.14	0.65	0.14	0.75	0.15
Beach	0.7	0.15	0.7	0.16	0.7	0.15	0.7	0.15
Buildings	0.85	0.20	0.8	0.18	0.75	0.15	0.65	0.15
Buses	0.95	0.19	0.95	0.19	0.95	0.19	0.9	0.19
Dinosaurs	1	0.20	1	0.20	1	0.20	1	0.2
Elephants	0.85	0.19	0.80	0.2	0.8	0.17	0.9	0.18
Flowers	0.95	0.19	0.95	0.2	0.95	0.19	0.95	0.2
Horses	1	0.18	0.95	0.19	0.95	0.19	1	0.2
Mountains	0.8	0.18	0.75	0.2	0.75	0.16	0.6	0.12
Food	0.8	0.18	0.8	0.2	0.85	0.18	0.65	0.14
Mean	0.86	0.179	0.845	0.186	0.835	0.173	0.82	0.168

Table 3. Average precision and recall for different methods using COIL 100

Method	Color space RGB		Color space HSV	
	Precision	Recall	Precision	Recall
Proposed method	0.95	0.245	0.97	0.25
[16]	0.947	0.217	0.955	0.235
[8]	0.939	0.210	0.949	0.225
[15]	0.917	0.182	0.925	0.20

space provides the best results of 0.97 for mean precision as shown in the proposed method, the result obtained for the RGB color space is 0.95. It can be clearly observed that proposed method is giving higher recall and precision rates as compared with other algorithms on Coil100 datasets.

4.3   Etude Du Temps CPU

The time taken by various methods for indexing and image search is shown in Table 4 for COIL-100 and Wang datasets. The time taken for feature extraction does not depend on image contents, it depends on the size of the image. It is shown in the table that the feature extraction time for [16], 0.284 s, is the long time and the time for [8], 0.185 s. The time taken for the proposed approach is 0.146 s which is less than the [15] approach which takes 0.250 s. The retrieval times are shown in columns three and four for COIL-100 and Wang datasets. Again, the image retrieval time on COIL-100 dataset for the proposed approach is 0.545 s, which is less than the time taken by Method [16] which is 1.350 s, which provides the best results. The retrieval time taken on Wang by the proposed method and Method [16] are 0.875 s and 2.305 s, respectively, which show that the proposed method provides faster image retrieval.

The total time taken for feature extraction and image retrieval for the proposed method which are 1.125 s and 1.545 s for COIL-100 and Wang are much less than that are required for Method [16], which are 1.94 s for COIL-100 and 2.905 s for Wang, Thus, the proposed method is faster than the two approaches whose retrieval performances are comparable to the performance of the proposed approach.

Table 4. Time taken (seconds) by various methods for image search

Approaches	Time taken for feature extraction (s)	Retrieval time(s)		Total time(s)	
		COIL-100	Wang	COIL-100	Wang
[16]	0.284	1.350	2.305	1.94	2.905
[8]	0.185	0.875	1.350	2.165	2.79
[15]	0.250	1.055	1.575	2.102	3.114
Proposed method	0.146	0.545	0.875	1.125	1.545

5   Conclusion

The prime contribution of this work is to build an efficient and effective CBIR system that tends to be feasible for large datasets. Therefore, this paper has presented an efficient image indexing and search system based on color and texture features. The color features are represented by combines 2-D histogram and statistical moments and texture features are represented by a gray level co-occurrence matrix (GLCM). A set of experiments was performed to choose the optimum vocabulary size that achieves the best retrieval performance. All considered retrieval procedures are examined on Wang and COIL-100 datasets in two color spaces HSV and RGB. The evaluation is carried out using the standard Precision and Recall measures. The results show that our proposed approach produces better results as compared to the existing methods. The proposed approach is effective in image retrieval. In future research, the proposed a smart color image retrieval scheme for combining all the three i.e. color, texture and shape information/feature, which achieved higher retrieval efficiency.

References

1. Flickner, M., Sawhney, H., Niblack, W., Ashley, J., Huang, Q., Dom, B., Gorkani, M., Hafner, J., Lee, D., Petkovic, D., et al.: Query by image and video content: the QBIC system. Computer **28**, 23–32 (1995)
2. Liu, Y., Zhang, D., Lu, G., Ma, W.Y.: A survey of content-based image retrieval with high-level semantics. Pattern Recogn. **40**(1), 262–282 (2007)
3. Chitaliya, N.G., Trivedi, A.I.: Comparative analysis using fast discrete Curvelet transform via wrapping and discrete Contourlet transform for feature extraction and recognition. In: 2013 International Conference on Intelligent Systems and Signal Processing (ISSP), pp. 154–159. IEEE, March 2013
4. Lei, Z., Fuzong, L., Bo, Z.: A CBIR method based on the color-spatial feature. In: TENCON 1999. Proceedings of the IEEE Region 10 Conference, vol. 1, pp. 166–169. IEEE (1999)
5. Zhang, D., Lu, G.: Shape-based image retrieval using generic Fourier descriptor. Signal Process.: Image Commun. **17**(10), 825–848 (2002)
6. Yang, M., Kpalma, K., Ronsin, J.: A survey of shape feature extraction techniques (2008)
7. Tilly, N.I.: Terrestrial laser scanning for crop monitoring. Capturing 3D data of plant height for estimating biomass at field scale, Doctoral dissertation, Universität ZU Köln (2015)
8. Youssef, S.M.: ICTEDCT-CBIR: integrating curvelet transform with enhanced dominant colors extraction and texture analysis for efficient content-based image retrieval. Comput. Electr. Eng. **38**(5), 1358–1376 (2012)
9. Rao, M.B., Rao, B.P., Govardhan, A.: CTDCIRS: content-based image retrieval system based on dominant color and texture features. Int. J. Comput. Appl. **18**(6), 40–46 (2011)
10. Yue, J., Li, Z., Liu, L., Fu, Z.: Content-based image retrieval using color and texture fused features. Math. Comput. Model. **54**, 1121–1127 (2011)
11. Ferreira, B., Rodrigues, J., Leitao, J., Domingos, H.: Practical privacy-preserving content-based retrieval in cloud image repositories. IEEE Trans. Cloud Comput. **PP**, 1 (2017)
12. Singha, M., Hemachandran, K.: Content-based image retrieval using color and texture. Signal Image Process. **3**(1), 39 (2012)
13. Xu, D., Xie, H., Yan, C.: Triple-bit quantization with asymmetric distance for image content security. Mach. Vis. Appl. **28**, 1–9 (2017)
14. Liu, D., Shen, J., Xia, Z., Sun, X.: A content-based image retrieval scheme using an encrypted difference histogram in cloud computing. Information **8**(3), 96 (2017)
15. Khalid, E.A., Chawki, Y., Aksasse, B., Ouanan, M.: Efficient use of texture and color features in content-based image retrieval (CBIR). Int. J. Appl. Math. Stat.™ **54**(2), 54–65 (2016)
16. Ashraf, R., Bashir, K., Irtaza, A., Mahmood, M.T.: Content-based image retrieval using embedded neural networks with palletized regions. Entropy **17**(6), 3552–3580 (2015)
17. Iqbal, Q., Aggarwal, J.K.: Combining structure, color, and texture for image retrieval: a performance evaluation. In: Proceedings of 16th International Conference on Pattern Recognition, 2002, vol. 2, pp. 438–443. IEEE.gov (2002)

18. Swain, M.J., Ballard, D.H.: Color indexing. Int. J. Comput. Vis. **7**(1), 11–32 (1991)
19. El Asnaoui, K., Aksasse, B., Ouanan, M.: Content-based color image retrieval based on the 2-D histogram and statistical moments. In: 2014 Second World Conference on Complex Systems (WCCS), pp. 653–656. IEEE, November 2014
20. Deegalla, S., Bostrom, H.: Reducing high-dimensional data by principal component analysis vs. random projection for nearest neighbor classification. In: 2006 5th International Conference on Machine Learning and Applications, ICMLA 2006, pp. 245–250. IEEE, December 2006
21. Wang, J.Z., Li, J., Wiederhold, G.: SIMPLIcity: semantics-sensitive integrated matching for picture libraries. IEEE Trans. Pattern Anal. Mach. Intell. **23**(9), 947–963 (2001)

Shape Classification Using B-spline Approximation and Contour Based Shape Descriptors

Youssef Ait Khouya[1(✉)] and Faouzi Ghorbel[2]

[1] Qualifying High School, Alhidaya Islamiya, Ministry of National Education,
Safi, Morocco
youssefaitkhouya@gmail.com
[2] GRIFT Research Group, CRISTAL Laboratory,
National School of Computer Sciences, University of Manouba, Manouba, Tunisia
faouzi.ghorbel@ensi.rnu.tn

Abstract. In this paper we propose to compare some of contour based shape descriptors like Fourier descriptors, Radial function and Fourier of Radial function after applying a global B-spline approximation to the contours. Applying global B-spline approximation we can smooth and reduce the number of points in the contours. We used this technique for classification of fossil species and especially Brachiopods. The fossils classification has a great importance in palaeontological studies. On the one hand, they make it possible to understand the biodiversity in its morphological dimension. On the other hand, they show the morphological transformations suffered during the biological evolution. We compared the descriptors using City block distance and the recognition rate to measure the discrimination of the descriptors.

Keywords: Shape classification · Brachiopods
B-spline approximation · Fourier descriptors · Radial function

1 Introduction

Shape descriptors is an important goal for image analysis and pattern recognition systems to enable the understanding of images [22]. A good number of shape descriptors, which are prevalent in the literature are largely categorized into two groups: contour based shape descriptors and region based shape descriptors. Contour-based shape descriptors use only boundary information, they cannot depict shape interior content. Of the contour based descriptors we cite Fourier descriptors [16, 24, 26, 31] which have been largely used. In [8] the authors propose a method to define Fourier descriptors for broken shapes, meaning shapes that can have more than one contour. Curvature approaches [18–20, 28] have also been used. In [18] a shape is described in a scale space by the maximum of the curvature. In [17] the authors presented a pattern description approach based on

© Springer International Publishing AG, part of Springer Nature 2018
A. Abraham et al. (Eds.): SoCPaR 2017, AISC 737, pp. 60–69, 2018.
https://doi.org/10.1007/978-3-319-76357-6_6

the multi-scale analysis of the contour of planar objects. Wavelet descriptors can be used to describe a given object shape by wavelet descriptors (WD) [27,29]. In region based techniques, all pixels surrounded by the shape boundary are taken into consideration to yield the shape descriptor. Frequently referred to methods are based on moment theory to describe shape [3,4]. These include geometric moments, Legendre moments, Zernike moments and pseudo Zernike moments.

In this paper we propose to compare some contour based shape descriptors like Fourier descriptor, Ghorbel descriptor, Radial descriptor and Fourier of Radial descriptor after applying a global B-spline approximation to the contours. The B-spline is largely used to describe a complex curve [11,12]. We applied this technique for Brachiopods classification. The fossils classification has a great importance in palaeontological field. There are many reasons for fossils classification such as: (i) We can remember the characteristics of a large number of different things only if they are grouped into categories whose members share given characteristics; (ii) To make better our predictive powers; (iii) Thanks to classification systems we are able to account for the ways apparently different things are related to each other. According for biologists the importance of the classification systems resides especially in reconstructing the evolutionary pathways that have produced the diversity of organisms living today. Several authors tried to classify fossil species. In [1] we presented a descriptor for Brachiopods classification it is a fusion between local and global complete and stable descriptors, and we cite [9], the authors used the Outline Shape Analysis technique for describing Ammonite shape. It aims at approaching the shape by a trigonometric function defined by the sum of sine and cosine terms. This latter is decomposed in either a series of harmonic amplitudes and phase angles or a series of Fourier coefficients, serving as variables for quantitative analysis. This technique was applied to other invertebrate groups such as Trilobites [10], bivalves [6], Ostracodes [2]. For edge detection we used the region-based active contour and especially the Chan and Vese model implemented by the Level set method [5]. The rest of the paper is organized as follows:

In Sect. 2, we remind the Region-based active Contour. Also, we formulate the Chan and Vese model in terms of level set functions and compute the associated Euler-Lagrange equation. In Sect. 3 we present the principle of B-spline approximation. The parametrization step are recaled in Sect. 4. The Fourier descriptor and Radial function are described in Sects. 5 and 6. Experimental results are given in Sect. 7, and we end the paper by a brief concluding section.

2   Overview of Region-Based Active Contour

The basic idea in active contour models or snakes is to evolve a curve, subject to constraints from a given image, in order to detect objects in that image. Starting with a curve around the object to be detected, the curve moves toward its interior normal and has to stop on the boundary of the object. Active contours based on region information are part of a very active area of research since the years 90 in Computer Vision. Originally, the work of [33] present a method of competitive regions in a mixed environment that is both Bayesian and minimizing the minimum description length criterion. In parallel, the work of Mumford and Shah realized in [21] have opened another branch, more geometric in the field of segmentation based on regions. On the other hand, many papers have dealt with region-based approaches using the level set framework, including the deformable regions by Jehan-Besson et al. [15] and the geodesic active regions by Paragios and Deriche [23], benefiting of adaptive topology at the expense of computational cost.

The region-based active contours model is based on the minimization of a well defined energy to segment the image. Let I an image and Ω his domain. Chan and Vese in [5] proposed to minimize the following energy:

$$E(c_1, c_2, \phi) = \mu \int_\Omega \mid \nabla H(\phi(x,y)) \mid dxdy \tag{1}$$

$$+ \nu \int_\Omega H(\phi(x,y))dxdy$$

$$+ \lambda_1 \int_\Omega \mid I(x,y) - c_1 \mid^2 H(\phi(x,y))dxdy$$

$$+ \lambda_2 \int_\Omega \mid I(x,y) - c_2 \mid^2 (1 - H(\phi(x,y)))dxdy$$

where $\mu \geq 0$, $\nu \geq 0$, $\lambda_1 > 0$, $\lambda_2 > 0$ are fixed parameters and H is the Heaviside function defined as:

$$H(z) = \begin{cases} 1 \text{ if } z \geq 0 \\ 0 \text{ if } z < 0 \end{cases}$$

and ϕ is the level set function associated to the contour evolving. The constants c_1 and c_2 are the averages of I in $\phi > 0$ and $\phi < 0$ respectively. So they are easily computed as:

$$c_1(\phi) = \frac{\int_\Omega I(x,y)H(\phi(x,y))dxdy}{\int_\Omega H(\phi(x,y)} \tag{2}$$

$$c_2(\phi) = \frac{\int_\Omega I(x,y)(1 - H(\phi(x,y)))dxdy}{\int_\Omega (1 - H(\phi(x,y))} \tag{3}$$

The discretized evolution equation is:

$$\frac{\phi_{(i,j)}^{n+1} - \phi_{(i,j)}^n}{\Delta t} = \delta_\epsilon(\phi_{(i,j)}^n)(\mu div(\frac{\nabla \phi_{(i,j)}^n}{\mid \nabla \phi_{(i,j)}^n \mid}) - \nu$$

$$-\lambda_1(I(i,j) - c_1(\phi_{(i,j)}^n))^2 + \lambda_2(I(i,j) - c_2(\phi_{(i,j)}^n))^2) \tag{4}$$

where $\delta_\epsilon(z)$ is a regular form of $\delta(z)$.

3   B-spline Approximation

B-splines are piecewise polynomial curves that are controlled by a set of points called the control points. This technique is widely used to represent a complex curve [11,12]. The B-spline curve is a generalization of the Bezier curve. Let $C(u)$ be the position vector along the curve as a function of the parameter u. The B-spline curve, $C(u)$, is defined as:

$$C(u) = \sum_{i=0}^{h} B_{i,p}(u)P_j \qquad (5)$$

where P_j is a control point, u is a parameter, and $B_{i,p}$ are the normalized B-spline basis functions of order p defined recursively as follows:

$$B_{(i,0)}(u) = \begin{cases} 1 \text{ if } u_i \leq u < u_{i+1} \\ 0 \text{ otherwise} \end{cases}$$
$$B_{(i,p)}(u) = \frac{u-u_i}{u_{i+p}-u_i} B_{(i,p-1)}(u) \qquad (6)$$
$$+ \frac{u_{i+p+1}-u}{u_{i+p+1}-u_{i+1}} B_{(i+1,p-1)}(u)$$

where u_i is known as a knot. In the approximation the B-spline curve does not have to pass through all data points except the first and last data points. A number of the B-spline control points would reflect the goodness of the approximation. Using least-square minimization, we compute the control points P_j, $j = 0, \ldots n$ of a B-spline curve $C(t)$ by minimizing the least-squares error defined as:

$$E(P_1, P_2, \ldots, P_n) = \sum_{k=1}^{m} |C(\overline{u_i}) - P_k| \qquad (7)$$

Where m is the number of data points and $\overline{u_i}$ is a parametric values of the points.

4   Parametrization

The contours Parametrization step provide all the computation results. For a given contour there are many types of parameter. A Parametrization function will be chosen which are suitable for the intended invariance. Here, we are interested in the parametrization invariant under the similarity transformation. To obtain a parametrization invariant under the similarity transformation we used the arc length parametrization, defined by the Euclidian norm which possesses an intrinsic link with the notion of similarity.

Suppose we have the coordinates of the contour $C(t) = (x(t), y(t))$, where t is a real value called the parameter of the contour. To reparametrize the contour by the arc length we need to calculate:

$$s(l) = \frac{1}{L} \int_0^l |C(t)'| \, dt \qquad (8)$$

where L denotes the curve length.

On the other hand, to obtain a parametrization invariant under affine transformation it's necessary to use an affine arc parametrization.

5   Overview of the Fourier Descriptors Methods

The Fourier Descriptor (FD) is a powerful tool for shape analysis and has been successfully applied to many shape representation applications. Let $x(t)$ and $y(t)$ be the coordinates of the contour, Since boundary is a closed curve and when the two-dimensional plane is considered as a complex plane, the cartesian coordinates of the contour are represented in the complex plane by $z(t) = x(t) + j.y(t)$. The discrete Fourier coefficients of $z(t)$ is defined as follow:

$$c_k = \sum_{k=0}^{N-1} z(t) exp(-j2\pi kt/N) \tag{9}$$

with $t \in [0, N-1]$, and N is a number of points in the contour.

Applying the inverse Fourier transform, we can recreate a contour from its Fourier descriptors. Initially, the first set of invariants was constructed by taking the modulus of Fourier descriptors.

$$\forall k \in [0, N-1], \quad I_k = |c_k|, \tag{10}$$

In [32] the authors used this method in their comparative study. The first coefficient is the only one depends on position of the shape. To obtain translation invariance the coefficient c_0 can be discarded. Scale invariance is achieved by dividing all the Fourier coefficients by the absolute value of the second coefficient. To obtain a rotation and shift invariant descriptor, the phase can completely ignored and the only use the absolute value of the Fourier coefficients.

The acquired Fourier descriptors is used to discriminate between simple-shaped objects [25,30]. But, this set of invariants is not complete in the sense defined by Crimmins in [7], A set of descriptors is said to be complete if the following property is verified:
two objects have the same shape if and only if they have the same set of invariants.

Completeness allows to retrieve shapes from their invariants, up to an Euclidean transformation. In fact, the set of descriptors defined in Eq. (10) is not complete since we can find objects with different shapes but with the same magnitude for their Fourier coefficients.

To resolve the problem of completeness, Crimins proposed in [7] a complete set of invariant, defined as:

$$
\begin{aligned}
I_{k_0} &= |c_{k_0}|, \quad for \;\; k_0 \;\; such \;\; that \;\; c_{k_0} \neq 0 \\
I_{k_1} &= |c_{k_1}|, \quad for \;\; k_1 \neq k_0 \;\; such \;\; that \;\; c_{k_1} \neq 0 \\
I_k &= c_k^{(k_0-k_1)} \; c_{k_0}^{(k_1-k)} \; c_{k_1}^{(k-k_0)}, \quad \forall k \neq k_0, k_1
\end{aligned}
\tag{11}
$$

However, this set is not stable, i.e. a slight modification of the invariants can induce a remarkable distortion of shape. To overcome this problem, Ghorbel in [13] presented a complete and stable set of invariant Fourier descriptors, defined as (the reader can consult Chap. II, pp. 53–71 in [14] for more details):

$$I_{k_0} = |c_{k_0}|, \quad for \quad k_0 \quad such \quad that \quad c_{k_0} \neq 0$$
$$I_{k_1} = |c_{k_1}|, \quad for \quad k_1 \neq k_0 \quad such \quad that \quad c_{k_1} \neq 0 \tag{12}$$
$$I_k = \frac{c_k^{(k_0-k_1)} \, c_{k_0}^{(k_1-k)} \, c_{k_1}^{(k-k_0)}}{|c_{k_1}|^{(k-k_0-p)} \, |c_{k_0}|^{(k_1-k-q)}}, \quad \forall k \neq k_0, k_1$$

with $p, q > 0$.

6   Radial Function

The Radial function is a function defined on a Euclidean space R^n that represents the distance from the digitized boundary of the object to its centroid as a function of distance along the boundary. The centroid point (x_c, y_c) is computed using the following formula:

$$x_c = \frac{1}{N} \sum_{i=1}^{N} x_i \tag{13}$$

$$y_c = \frac{1}{N} \sum_{i=1}^{N} y_i \tag{14}$$

where (x_i, y_i) is the discreet contour and N is the number of the boundary points. The Radial function $\rho(i), i = 1, 2, \ldots N$, are calculated starting from an fixed position of the boundary using as distance measure on a Euclidean norm from the centroid to the boundary points as:

$$\rho(i) = \sqrt{(x_i - x_c)^2 + (y_i - y_c)^2} \tag{15}$$

Since the radial function is only dependent on the location of the centroid and the points on the boundary, it is invariant to the translation due to the subtraction of the centroid from boundary coordinates, and if the starting point is fixed the function is invariant to the rotation, then the function of the original and rotated shapes will be identical. This function alone is not invariant to a change in starting point or scaling. In order to obtain a descriptor that is invariant to translation, rotation, scaling, and change of starting point we apply the Fourier transformation, and then take the magnitude of these normalized coefficients defined as:

$$V = \left(\frac{|F\rho(1)|}{|F\rho(0)|}, \frac{|F\rho(2)|}{|F\rho(0)|}, \ldots, \frac{|F\rho(N)|}{|F\rho(0)|} \right) \tag{16}$$

where F is the discrete one-dimensional Fourier transform.

7   Experimental Results

Following the preprocessing stage, the contour is detected using region based active contours method, specifically chan and vese model implemented with the

Level Set method, this model provides satisfactory results for the contour detection in an image (see Sect. 2). Every object is represented by the x and y coordinates of its boundary points. The number of these points varies from 600 to 1200 for images in our databases. We have re-sampled the contour in $N = 2^{10} = 1024$ points. When the pretreatment step is completed, we approximate the contours using global B-spline approximation, in order to reduce the number of points in the contour. Many spline function are available in the literature but here we use a cubic B-spline because it'is the most popular in engineering application and also is a smoothest function, is used to smooth the contour points to reduce the noise effect which can be product in edge detection step.

We realized a comparative study of Ghorbel descriptor, Radial function and Fourier of Radial function for classification of Brachiopods. We dispose a database of Brachiopods containing 20 classes and each class contains 20 images. Samples of the Brachiopods database are depicted in Fig. 1.

For two shapes represented by their descriptors, the similarity between the two shapes is measured by the city-block distance between the two feature vectors of the shapes. Therefore, the online matching is efficient and simple. We used the recognition rate to compare the discrimination of the descriptors. For efficient shape description, only a small number of feature are used. We applied request the seven groups each group contains 100 images of Brachiopods. The first two

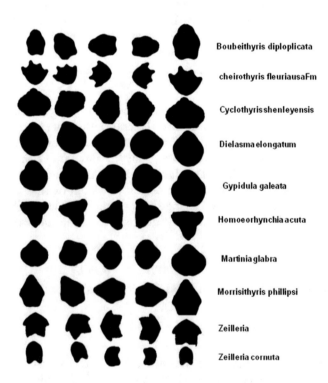

Boubeithyris diploplicata

cheirothyris fleuriausaFm

Cyclothyris shenleyensis

Dielasma elongatum

Gypidula galeata

Homoeorhynchia acuta

Martinia glabra

Morrisithyris phillipsi

Zeilleria

Zeilleria cornuta

Fig. 1. Samples of the Brachiopods database.

groups contain images of the database. Others contain images that do not belong to the database, they are the images of database that have met a change as rotation and scale. The results of the comparison are shown in Fig. 2. The Table 1 present a comparison of the run time of the three descriptors executed on a processor machine Intel Core 2 Duo 2 GHz with 2 GB of RAM.

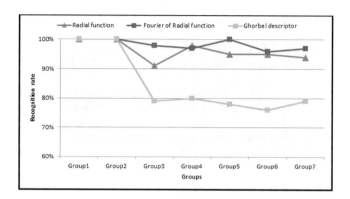

Fig. 2. Recognition rate of Brachiopods groups.

Table 1. Comparison of the run time of the three descriptors

Descriptors	Run time (ms)
Radial function	0.031
Fourier of radial function	0.056
Fourier descriptors (Ghorbel)	0.444

8   Conclusions

The comparison of some contour based shape descriptors like Fourier descriptors, Radial function and Fourier of Radial function after applying a global B-spline approximation to the contours is proposed in this paper. The raisons for applying a global B-spline approximation are: We can smooth and reduce the number of points in the contours. We used the recognition rate to measure the discrimination of the descriptors and the City block distance as a classifier. We applied this technique for Brachiopods classification. The results of this comparison show that the Fourier of radial function gives the greatest recognition rate followed of the radial function and ultimately the Fourier descriptors.

References

1. Khouya, Y.A., Ghorbel, F., Alaa, N.: Brachiopods classification based on fusion of global and local complete and stable descriptors. In: 12th ACS/IEEE International Conference on Computer Systems and Applications, Marrakech, Morocco, 17–20 November 2015
2. Bachnou, A.: Modélisation des contours fermés: an-morph outil mis au point pour maitriser les composantes des profils latraux des ostracodes. perspective d'application systmatique. Géobis **32**(5), 733–742 (1999)
3. Bailey, R., Srinath, M.: Orthogonal moment features for use with parametric and non-parametric classifiers. IEEE Trans. Pattern Anal. Mach. Intell. **18**(4), 389–399 (1996)
4. Belkasim, S., Shridar, M., Ahmadi, M.: Orthogonal moment features for use with parametric and non-parametric classifiers. IEEE Trans. Pattern Anal. Mach. Intell. **24**, 1117–1138 (1991)
5. Chan, T., Vese, L.: Active contours without edges. IEEE Trans. Image Process. **10**(2), 266–277 (2001)
6. Crampton, J.: Elliptic Fourier shape analysis of fossil bivalves: some practical considerations. Lethaia **28**(2), 179–186 (1995)
7. Crimmins, T.R.: A complete set of Fourier descriptors for two dimensional shapes. IEEE Trans. Syst. Man Cybern. **12**, 848–855 (1982)
8. Dalitz, C., Brandt, C., Goebbels, S., Kolanus, D.: Fourier descriptors for broken shapes. EURASIP J. Adv. Sig. Process. **2013**, 161 (2013)
9. El Hariri, K., Bachnou, A.: Describing ammonite shape using Fourier analysis. J. Afr. Earth Sci. **39**, 347–352 (2004)
10. Foote, M.: Perimeter-based Fourier analysis: a new morphometric method applied to the trolobite cranidium. J. Paleontol. **63**(6), 880–885 (1989)
11. Cohen, F.S., Huang, Z., Yang, Z.: Invariant matching and identification of curves using B-splines curve representation. IEEE Trans. Image Process. **4**(1), 1–10 (1995)
12. Farin, G.E.: Curves and Surfaces for CAGD: A Practical Guide, 5th edn. Morgan Kaufmann, Burlington (2001)
13. Ghorbel, F.: Stability of invariant Fourier descriptors and its inference in the shape classification. In: 11th International Conference in Pattern Recognition, The Hague, The Netherlands, 30 August–3 September 1992
14. Ghorbel, F.: Invariants de formes et de mouvement Onze cas du 1D au 4D et de l'euclidien aux projectifs. Arts Pi, Tunis (2013)
15. Jehan-Besson, S., Barlaud, M., Aubert, G.: DREAM^{2}S: deformable regions driven by an Eulerian accurate minimization method for image and video segmentation. Int. J. Comput. Vis. **53**(1), 45–70 (2003)
16. Kauppinen, H., Seppnen, T., Pietikinen, M.: An experimental comparison of autoregressive and Fourier-based descriptors in 2D shape classification. IEEE Trans. Pattern Anal. Mach. Intell. **17**(2), 201–207 (1995)
17. Kpalma, K., Ronsin, J.: Multiscale contour description for pattern recognition. Pattern Recogn. Lett. **27**(13), 1545–1559 (2006)
18. Mokhtarian, F., Abbasi, S.: Shape similarity retrieval under affine transforms. Pattern Recogn. **10**(2), 31–41 (2002)
19. Mokhtarian, F., Mackworth, A.K.: Scale-based description and recognition of planar curves and two-dimensional shapes. IEEE Trans. Pattern Anal. Mach. Intell. **8**(1), 34–43 (1986)

20. Mokhtarian, F., Mackworth, A.K.: A theory of multiscale, curvaturebased shap representation for planar curves. IEEE PAMI **14**, 789–805 (1992)
21. Mumford, D., Shah, J.: Optimal approximation by piecewise smooth functions and associated variational problems. Commun. Pure Appl. Math. **42**, 577–685 (1989)
22. Nagabhushana, S.: Computer Vision and Image Processing. New Age International (P) Limited, Delhi (2005)
23. Paragios, N., Deriche, R.: Geodesic active regions and level set methods for supervised texture segmentation. Int. J. Comput. Vis. **46**(3), 223–247 (2002)
24. Persoon, E., Fu, K.: Shape discrimination using Fourier descriptors. IEEE Trans. Syst. Man Cybern. **7**(3), 170–179 (1977)
25. Persoon, E., Fu, K.: Shape discrimination using Fourier descriptors. IEEE Trans. Pattern Anal. Mach. Intell. **8**(3), 388–397 (1986)
26. Rui, Y., She, A.: A modified Fourier descriptor for shape matching in mars. Image Databases Multimed. Search **8**, 165–180 (1998)
27. Tieng, Q.M., Boles, W.W.: Recognition of 2D object contours using the wavelet transform zero crossing representation. IEEE Trans. PAMI **19**(8), 910–916 (1997)
28. Urdiales, C., Bandera, A., Sandoval, F.: Non-parametric planar shape representation based on adaptive curvature functions. Pattern Recogn. **35**, 43–53 (2002)
29. Yang, H.S., Lee, S.U., Lee, K.M.: Recognition of 2D object contours using starting-point-independent wavelet coefficient matching. J. Vis. Commun. Image Represent. **9**(2), 171–181 (1998)
30. Zahn, C., Roskies, R.: Fourier descriptors for plane closed curves. Trans. Comput. **21**(3), 269–281 (1972)
31. Zhang, D.S., Lu, G.: A comparative study of Fourier descriptors for shape representation and retrieval. In: Proceedings of 5th Asian Conference on Computer Vision (ACCV 2002), 22–25 January, pp. 646–651, Melbourne, Australia (2002)
32. Zhang, D., Lu, G.: Study and evaluation of different Fourier methods for image retrieval. Image Vis. Comput. **123**, 33–49 (2005)
33. Zhu, S., Yull, A.: Region competition: unifying snakes, region growing, and Bayes/MDL for multiband image segmentation. IEEE Trans. Pattern Anal. Mach. Intell. **18**, 884–900 (1996)

An Image Processing Based Framework Using Crowdsourcing for a Successful Suspect Investigation

Hasna El Alaoui El Abdallaoui$^{(\boxtimes)}$, Fatima Zohra Ennaji,
and Abdelaziz El Fazziki

Computing Systems Engineering Laboratory, Cadi Ayyad University,
Marrakesh, Morocco
h.elalaoui@edu.uca.ac.ma, f.ennaji@edu.uca.ma,
elfazziki@uca.ma

Abstract. The invasion of new technologies in people's life has allowed a great interactive collaboration between citizens and law enforcement agencies. The appearance of crowdsourcing has become a new source of research and development especially in the suspect investigation domain that needs the combination of human intelligence and the technical tools to lead the investigation towards the greatest results. The objective of this paper is to exploit the pervasiveness of image processing techniques (face detection and recognition) to design a crowdsourcing framework that may be chiefly used by government authorities to identify a suspect. This framework is primarily based on the surveillance video analysis and the sketch generation tools supported by the intelligence of the crowd.

Keywords: Crowdsourcing · Image processing
Face detection and recognition · Suspect investigation

1 Introduction

Human intelligence has been the key to all the technological advances that humanity is experiencing. With the advent of the Internet and mobile devices, human intelligence has taken on a new form in our actual societies and has, however, enabled a better supply and processing of information. This has been conceptualized as 'crowdsourcing' and has been officially coined by Howe [1].

Crowdsourcing is defined as participatory production based on a collaborative logic, users use the creativity, intelligence and know-how of each one to answer a given problem or to perform a task. Since its emergence, crowdsourcing has contributed in several areas and more governmental organizations have integrated it into their decision-making policy [2]. Crowdsourcing initiatives typically focus on citizen online mobility as a new resource for innovation and problem solving for government agencies: citizens are no longer claimants of rights but contribute in several situations such as the identification of suspects.

Although crowdsourcing is increasingly being used by many organizations, some tasks are far from being feasible for human beings, either because they involve a lot of

© Springer International Publishing AG, part of Springer Nature 2018
A. Abraham et al. (Eds.): SoCPaR 2017, AISC 737, pp. 70–80, 2018.
https://doi.org/10.1007/978-3-319-76357-6_7

resources, require considerable computing time, or because they need more sophisticated tools for their achievement. Image processing is one of the disciplines that marks
this exception and has, however, incorporated a set of very advanced algorithms and
methods enabling humans to better study the image and its transformations. The
conditions of brightness, lightening and color [3] are still major constraints in the image
processing and more particularly in the facial detection and recognition [4] that are
widely used in law enforcement.

This paper aims to exploit the crowdsourcing potential to obtain more accurate
values by supporting or correcting the results of image processing techniques. For this
purpose, a framework is proposed which allows the identification of a suspect from
surveillance cameras or the suspect identikit generated from the personal description
provided by the crowd.

The rest of the paper is organized as follows: the Sect. 2 presents a state of art related
to the crowdsourcing concept and the suspect investigation. In Sect. 3 the approach of
the proposed framework and its architecture will be described. A case study was handled
in Sect. 4 while Sect. 5 discusses the results in a conclusion.

2   State of the Art

In this section, we present some useful information in the form of a synthesis in the two
areas of interest of this paper, namely crowdsourcing in e-governmental applications and
suspect investigations that have improved thanks to technological and scientific advances.

2.1   Crowdsourcing

The emergence of crowdsourcing. The term "crowdsourcing" was first coined by Jeff
Howe and Mark Robinson in a Wired magazine article [1] as a fusion of two words:
'crowd' and 'outsourcing'. It was defined as the outsourcing of a task to a person or a
group of people in a form of an open call [1]. It is important to highlight that the call
should not be limited to experts or selected participants. Crowdsourcing is often based
on the idea of collective intelligence, the aim is to make the knowledge the most
accurate by giving contributions from a distributed population – "all of us together are
smarter than any one of us individually…" [5]. Since its creation, crowdsourcing has
become the subject of several research and studies.

Crowdsourcing and E-government. Crowdsourcing became a tool used for a wide
group of governance activities to help government to engage with citizens: when a
decision maker has a problem to accomplish some activities, he may involve a
crowdsourced participation. Through a crowdsourcing platform, the decision maker
(the crowdsourcer) submits the task and the crowdsourcing platform transmits it to the
crowd. The appropriate participants are chosen depending on the crowdsourcing
activity. Once crowdworkers finish the task, they send the results via the same platform
to the crowdsourcer who evaluates the received data and selects the satisfying results.
In some crowdsourcing activities, incentives and rewards may be offered to the participants. The crowdsourcers and crowdworkers may have any direct communications,
through email, telephone, etc.

2.2 Suspect Investigation

The most comprehensive, practical and reliable manual on criminal investigation have been developed in [6]. A definition of criminal investigation has been proposed as a process of: *"...Discovering, collecting, preparing, identifying and presenting evidence [before a tribunal of fact] to determine what happened and who is responsible"*.

The suspect investigation requires that a series of steps must be taken in the proper sequence. A suspect investigation is usually initiated by an observation or a reporting of a crime. Then follows a process that may include all of some of these steps [7]:

- Initial investigation: it identifies important and informative elements through essential methods. It emphazes two procedures: Collecting and preserving of evidences and collection and interpretation of testimonies [7].
- Documentation: consists of carrying out a complete and detailed synthesis of all available information.
- Follow-up investigation: based on two steps. The first one is the reconstruction that is gathering the results of all collected data and secondly, the identification of the suspect. In this paper, we will mainly focus on the 'personal description' method to identify the suspect, the most important among eight other techniques [7].
- Prosecution and arrest: is the pursuit of any trace that may lead to the suspect arrest.

2.3 Forensic Investigation

The same authors in [6] differentiated between the suspect and forensic investigation; which means the introduction of technological methods and scientific processes in a criminal investigation [6]. The forensic methodology extracts evidences from digital platforms or exploit them to identify or analyze digital evidence. Hereafter, two of these digital techniques heavily used in law enforcement.

Video-based surveillance systems. Governments, public companies, private sector companies and individuals spend a good budget on surveillance cameras in order to secure their properties, prevent crimes and apprehend criminals. During the past decade, video surveillance systems have grown thanks to advances in computer vision research. The proof is the worldwide, continuing demand for video surveillance systems: The world market for video surveillance forecasts to increase by 7% in 2017. In the same year, 28 million high-definition cameras will be marketed against 66 million IP cameras [8]. Despite the fact that many problems (object detection and recognition, occlusion problems, etc.), are still far from being perfectly solved, video surveillance systems are becoming increasingly intelligent. From analogue CCTV systems, to fully digitized and automated network-based video surveillance, these systems experienced a great technological evolution [9, 10].

Performing digital identikits (or facial sketches). Sketches are widely used in criminal investigations to help in the identification and apprehension of criminals. They are carried out by means of a verbal description provided by the victims and witnesses and are then distributed in the media or posted in public places for possible identification by the crowd [11]. Thanks to technological evolution, they're making moves from free hand sketching (forensic sketch) drawn by forensic artists to computer kits (composite sketches) [12].

3   The Proposed Approach

It is in this section of the paper that the proposed approach will be described. As a first step, we define the problem that this solution tries to solve and then suggest a scenario that details the steps of the process. In a second part, we design the architecture of the framework in addition to all the components and their implementation.

3.1   Problem Definition

New technologies offer very advanced functionalities and possibility for human being to carry out tasks hitherto difficult to complete by himself. Intelligent surveillance cameras and sketch generation tools are among the technological achievements that have marked the crime-solving domain. Crowdsourcing, has become a model of aid and support that is widely used in various domains including law enforcement where crowdsourcing was the culmination in the resolution of several cases. The principle is to appeal the human intelligence for the search or the description of a suspect. Within this schema, we propose a framework that will link human intelligence and techno-logical advances based primarily on the techniques of automatic face analysis in sus-pect investigation. This framework aims to facilitate the interaction between the crowd and the police in order to collect, analyze and share useful information to bring the investigation to successful results. All the functionalities and the process that the framework we are proposing follows will be detailed in the section below.

3.2   The Suspect Identification Process

To describe the process followed by the framework (Fig. 1), we first suppose that a suspect committed a criminal act (aggression, an armed robbery, etc.) and then fled. The authorities, then, check the surveillance cameras surrounding the crime's place. If a video is available, then starts the video analysis stage to retrieve the clearest picture of the perpetrator. If the police obtains a clear-cut image of the suspect, the automatic face analysis techniques will be the best way to search for a matching between the suspect that the police is searching for and a former criminal previously registered in the police database. If a matching exists, the crowd ensure the identity of the suspect. The identified suspect is then regarded as a criminal with judicial case so his profile is shared with the public to start the search process. In the case where no matching is found, the picture retrieved from the surveillance video is disseminated. Contrariwise, i.e. if no video has been recorded or the video is heavily unclear, the personal description is a prominent key in the identification of the suspect. Thanks to the police expertise in interviewing the victims and eyewitnesses, they can obtain valuable information concerning the suspect profile. This helps to perform the facial identikit of the suspect that will be spread once approved by the interviewees.

The next crowdsourced task is to alert the police about the suspect's whereabouts. Nowadays, the crowd is always accompanied by mobile devices equipped with rich sensors enabling them to share real-time location data.

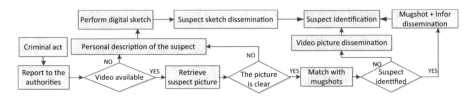

Fig. 1. The suspect identification process

3.3   The Proposed Approach Architecture

The framework we are proposing is a crowdsourcing application where several technical modules were incorporated and that are essentially based on facial detection and recognition methods. From the foregoing, the framework architecture is presented in Fig. 2 hereafter and each component is described separately.

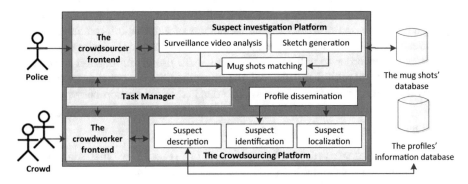

Fig. 2. The proposed approach architecture

The architecture's components can be described as follows:

- The crowdsourcer frontend: it is a Human-Machine Interface (HMI) that lets the police to control all the police investigation procedures. Thanks to this interface, all operations of the facial analysis will be carried out; this includes the analysis of surveillance videos and the generation of suspects' identikits. For any result obtained at the end of these operations, a matching with the profiles of former criminals will be performed for a possible identification of the suspect.
- The crowd frontend: it is the bridge that allows participants to access the crowdsourcing platform and permits them to perform all tasks assigned to them.
- Profile dissemination: is a common component between the police and the crowd. Its purpose is to widely disseminate the suspect profile (with the crowd and between the crowd's networks) in order to speed up the suspect identification.
- The task manager: As its name implies, this component is responsible for managing tasks between the crowdsourcer and the crowdworker. It defines the specifications of the crowdsourced tasks and their allocations (execution, participants, etc.).

Hereafter, a comprehensive description of the two platforms that constitute the contribution of this paper.

The Suspect Investigation Platform. It is the platform where all the image processing procedures occur. It integrates all the necessary modules:

Surveillance video analysis. Actually, surveillance cameras are real-time alerting systems. Thanks to image/video processing techniques, they are experiencing a continuous development. To do this, several methods of automatic face analysis will be exploited. In general, two main steps must be implemented:

- Face detection: It aims to distinguish a human face in an image. It is an indispensable phase in the process of facial recognition. The most known method is the Viola-Jones algorithm [13]. This method quickly became a standard in the field of computer vision since it incorporated new concepts and methods: The "Integral image" which is a notable gain in terms of calculating the HAAR characteristics. In the detection phase, the whole image is traversed by moving the detection window of a certain step in the horizontal and vertical directions. The second key element of the Viola and Jones method is the use of a boosting method to select the best features. The boosting algorithm used is in practice a modified version of AdaBoost [14]. One of the key ideas of the method to reduce the computational cost is to organize the detection algorithm into a cascade of classifiers. For more clarity, we have summarized the key steps of the Viola-Jones algorithm in Fig. 3.

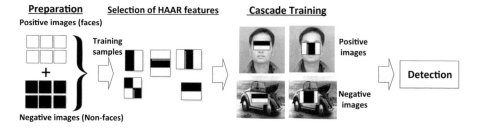

Fig. 3. The Viola-Jones algorithm

- Face recognition: is a field of computer vision consisting of automatically recognizing a person from an image of his/her face. Facial recognition algorithms extract the facial features and compare them to a database to find the best match. In this paper, we present the FisherFaces recognition method [15] based on two concepts: the EigenFace algorithm (PCA) and the Linear Discrimination Analysis (LDA). The first system of face recognition that yielded significant results was achieved using the so-called "Eigenfaces" method [16]. The "Fisherfaces" method has been proposed to solve the problem of robustness in the face of variations in pose, illumination and expressions that challenge the PCA. The Fig. 4 below presents the steps of the FisherFaces algorithm.

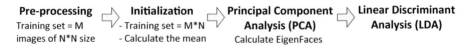

Fig. 4. The FisherFaces method steps

Identikit generation. Thanks to the description given by the crowd, the forensic artist will draw the suspect's sketch based on the suspect characteristics. The generation of identikits moves from free hand drawing to the use of recent technological tools. Police draughts men take special training to make sketches. To perform sketches, we used in this framework Faces IQ Biometrix software tool. It is used by worldwide police agencies because of its ease of use and its large facial features database.

The Crowdsourcing Platform. In the proposed framework, three major tasks will be the crowdsourcing activities delegated to the crowd. The first one is personal description, where the crowd is asked to provide the police with any information about the suspect (physical, vocal characteristic, etc.) using interview techniques [7]. The personal description is one of the main keys for a successful identification in the suspect investigation. In this framework, we did not use all the physical properties, but we were interested only on the facial attributes in addition to the gender and the skin color (see Table 1). These information are the most prominent for the sketch generation. The second task is the suspect identification that aims to confirm, directly or indirectly, the identity of the suspect by the crowd. The last task given to the crowd is the suspect localization. After the police disseminated the suspect's profile, the crowd has to report any trace of the suspect once located.

Table 1. Some examples of facial attributes for personal description

Facial attribute	Property
Gender	Male – Female
Human race	Black – White – Yellow – Red – Dark
Face shape	Oval – Round – Heart – Long – Triangle
Hair color	Black – Brown – Blond – RedHead – White
Hair length	Extra-long – Medium – Short – Extra short
Eyebrows	Soft – Thick – Unibrow – Straight – High/low arch
Eyes shape	Buried - Bent – Sharp – Small – Baggy – Round – Line
Nose shape	Pointed – Broken – Long – Droopy – Flat – Bulbous – Grecian
Mouth shape	Thin lips – Pulpy – Small – Big lips – Dissimilar lips

4 Case Study

A simulation of a police investigation has been done to test the effectiveness of our framework. The human intelligence cannot be compared with the machine one, more precisely in the facial analysis area. Whence comes the proposed solution that integrates the machine and human intelligence to contribute on the security of our country.

As a face recognition database, we used the MIT CBCL Face Data Set [17] provided by the Center for Biological and Computational Learning at MIT. All the people that exist in this database were considered as former criminals. Their images have been stored in addition to some supplementary information. Figure 5 shows a Java interface containing the suspects' information and their mug shots. In the following, we are going to give some examples of both cases.

Fig. 5. A part of the criminals' database

4.1 Surveillance Video Analysis

OpenCV image-processing library was used in our application. In most cases, the suspect is detected using the face detection and recognition algorithms explained in the previous section. However, the machine remains helpless and may give wrong answers like as it is illustrated in Fig. 6. The poor interpretation comes from several facts such as the quality of the video, the invisible parts of the face, etc. In such situation, crowd tasks are created to extract the maximum of personal information to construct a profile close to the suspect's one. The crowd interrogated had only a few seconds to visualize the face (the 4th face in Fig. 5) in all possible positions without knowing that they must provide details on each facial attribute. Table 2 shows the actual values and the percentage of people who responded correctly. The data provided from the crowd were closer to the actual values and allowed more precise personal description than that delivered by the algorithms of facial detection and recognition.

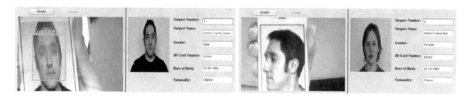

Fig. 6. A successful and a failed case in suspect identification from a real time video

Table 2. Some of the crowd results for personal description

Facial attribute	Gender	Age	Human race	Face shape	Hair color	Hair length
Real value	Male	20–30	White	Oval	Brown	Short
Answers	100%	51.2%	81.4%	30.2%	79.1%	74.4%
Facial attribute	Hair density	Eyebrows	Eyes shape	Eyes color	Nose shape	Mouth shape
Real value	Heavy	Thick	Round	Green	Pointed	Small
Answers	55.8%	55.8%	18.6%	23.3%	25.6%	30.2%

4.2   The Generated Sketch

The second experiment aims to simulate a digital sketch recognition. To do so, two faces were randomly chosen from the MIT face database to elaborate their sketch in addition of a real criminal image with its corresponding sketch (see Fig. 7).

Fig. 7. Two MIT faces and their corresponding sketches & a real criminal mug shot and sketch

As the previous experiment, the results show that the first and the last case were successful since it gives the corresponding image of each sketch (Fig. 8). Conversely, the second case when the quality of the sketch has been manipulated, our application couldn't figure out the owner of the sketch (see Fig. 8). This problem might happen either when the sketch is not well done or doesn't reflect the real face features of the suspect. Once again, a task is created making a use of the crowd intelligence to overcome this issue. The crowd participants were asked to recognize the owner of the sketch from four images under different positions. The results (in Fig. 9) show that the majority of the crowd (60.5%) have identified the right answer (suspect 3).

Fig. 8. Two successful sketch automatic recognitions and a fail case due to bad-quality sketch

Fig. 9. Crowd sketch recognition results

5   Discussion and Conclusions

Crowdsourcing gradually enters our society and tends to change some of our ways of thinking or acting. It creates a phenomenon of community and popular collaboration to perform various tasks. But crowdsourcing would never have gained so much extent without the enormous capabilities offered by new technologies.

In this paper, we concluded that crowdsourcing can be a great support for governmental organizations to access citizens as a source of information and particularly in suspect investigation where the accuracy of data is very important. The identification of suspects requires a good execution of the image processing methods, in particular the techniques of face detection and recognition applied to surveillance videos and identikits generation.

However, we have proved that the fusion of human intelligence and technological advances allows better analysis of data and more precise results and thus a success in the investigation and identification of suspects.

References

1. Howe, B.J.: The rise of crowdsourcing. Wired Mag. **14**(6), 1–4 (2006)
2. Doan, B.Y.A., Ramakrishnan, R., Halevy, A.Y.: Crowdsourcing systems on the world-wide web. Commun. ACM **54**(4), 86–96 (2011)
3. Yang, M.H., Kriegman, D.J., Ahuja, N.: Detecting faces in images: a survey. IEEE Trans. Pattern Anal. Mach. Intell. **24**(1), 34–58 (2002)
4. Ahonen, T., Hadid, A., Pietikäinen, M.: Face description with local binary patterns: application to face recognition. IEEE Trans. Pattern Anal. Mach. Intell. **28**(12), 2037–2041 (2006)
5. Aitamurto, T., Leiponen, A., Tee, R.: The promise of idea crowdsourcing – benefits, contexts, limitations. Nokia IdeasProject White paper, vol. 1, pp. 1–30 (2011)
6. Hess, K.M., Orthmann, C.H.: Criminal Investigation, 9th edn. Cengage Learning, Boston (2016)
7. Sefanyetso, J.T.: Personal description: an investigation technique to identify suspects (2009)
8. Les Grands Tendances 2017 De La Vidéo Surveillance. http://www.video-surveillance-direct.com/content/88-2017-les-tendances-de-la-video-surveillance. Accessed 05 July 2017
9. Zhang, Z.: Intelligent distributed surveillance systems: a review. Chin. J. Electron. **14**(1), 115–118 (2005)
10. Swanson, C., Chamelin, N., Territo, L., Taylor, R.: Criminal Investigation (The Evolution of Criminal Investigation and Forensic Science). The Granger Collection, New York (2011)
11. Klum, S., Han, H., Jain, A.K., Klare, B., Church, F.: Sketch based face recognition : forensic vs composite sketches. In: International Conference on Biometrics, pp. 1–8. IEEE, Madrid (2013)
12. Streed, M.W.: Creating Digital Faces for Law Enforcement, 1st edn. Academic Press, Cambridge (2017)
13. Viola, P., Jones, M.J.: Robust real-time face detection. Int. J. Comput. Vis. **57**(2), 137–154 (2004)
14. Freund, Y., Schapire, R., Abe, N.: A short introduction to boosting. J. Japan. Soc. Artif. Intell. **14**(5), 771–780 (1999)

15. Hjelmas, E., Low, B.K.: Face detection: a survey. Comput. Vis. Image Underst. **83**(3), 236–274 (2001)
16. Turk, M.A., Pentland, A.P.: Face recognition using eigenfaces. In: Computer Society Conference on Computer Vision and Pattern Recognition, pp. 586–591. IEEE (1991)
17. Weyrauch, B., Heisele, B., Huang, J., Blanz, V.: Component-based face recognition with 3D morphable models. In: IEEE Computer Society Conference on Computer Vision and Pattern Recognition Workshops. IEEE (2004)

Recognition and Classification of Arabic Fricative Consonants

Youssef Elfahm[1]([⊠]), Badia Mounir[2], Ilham Mounir[2],
Laila Elmaazouzi[2], and Abdelmajid Farchi[1]

[1] IMII Laboratory, Faculty of Sciences and Technics, Settat, Morocco
y.elfahm@gmail.com, abdelmajid.farchil@gmail.com
[2] LAPSSII Laboratory, Graduate School of Technology, Safi, Morocco
mounirbadia2014@gmail.com, ilhamounir@gmail.com,
Elmazouzi2001@yahoo.fr

Abstract. The aim of this paper is to develop and study the performances of a system of recognition and classification of Arabic fricative consonants. We developed under Matlab an algorithm that allows the calculation and the extraction of acoustic cues dependent on the energy distribution in the Syllables CV where C refers the Arabic fricative consonants and V refers to one of the three vowels /a/, /u/ or /i/. We then used the decision trees (J48) to classify these syllables. The obtained results show that this system allows a good classification of these consonants.

Keywords: Automatic speech recognition · Classification · Acoustic cues
Fricative consonants · Decision trees

1 Introduction

The study of properties allowing the identification and the classification of speech sounds is the major objective of speech processing researchers. Several studies were interested in the identification and classification of acoustic signals (fricatives, stops, nasals,…). However, the task is complicated by the problem of variability in the acoustic signal.

Fricatives are produced with a very narrow constriction in the oral cavity. A rapid flow of air through the constriction creates turbulence in the flow, and the random velocity fluctuations in the flow can act as a source of sound. Air turbulence produced in this way, by various kinds of constrictions in the vocal tract (the position of which depends on the particular fricative), is the typical sound source for all fricatives [1, 2].

Many studies in English voiceless fricatives, reported that non-sibilants (/f, h/) have low amplitude than sibilants (/s, S /), with no differences within each class [3, 4]. Shadle reported that the increase in sibilance amplitude is due to increased turbulence in the airflow caused by the lower teeth acting as an obstacle to the noise source of the constriction sibilant [2].

The examination of spectral moments of English fricatives can differentiate /S / from /s/ in terms of spectral mean [5–7], kurtosis, skewness [8] and standard deviation [9]. Tomiak reported that /h/ displays a greater standard deviation, skewness, and

© Springer International Publishing AG, part of Springer Nature 2018
A. Abraham et al. (Eds.): SoCPaR 2017, AISC 737, pp. 81–89, 2018.
https://doi.org/10.1007/978-3-319-76357-6_8

kurtosis than /f/ [9]. Shadle and Mair have shown that / S / is among the English fricatives that have a lower spectral average [10]. Jongman et al. reported that variance is high for the non-sibilants and low for the sibilant fricatives. He also reported that skewness and spectral peak location help distinguishing English fricatives in terms of place of articulation [7].

More recently, according to ElinaNirgianak, the normalized amplitude differenti-ated sibilants from non-sibilants, F2 onset and spectral mean are the parameters that distinguished all five places of articulation. Also, she has reported that normalized duration and normalized amplitude are the parameters that distinguish Greek voiced from voiceless fricatives, with important classification accuracy [11].

Few studies have been interested on Arabic fricative characterization, Al-khairy has reported that the overall absolute frication noise duration of sibilant fricatives (mean 138.09 ms) was longer than non-sibilants (mean 109.34 ms). Moreover, he has noticed that frication noise duration of voiceless fricatives (mean 134.21 ms) was longer on average than that of voiced fricatives (mean 92.05 ms). Furthermore, Al-khairy has reported that amplitude measurements differentiated sibilant fricatives (/s, sç, z, ∫/) as a class from non-sibilants (/f, θ, ð, ðç/) while failing to distinguish within each of the two classes [12].

Concerning spectral peak location of fricatives, the results of Al-khairy reported that spectral peak location distinguished non-sibilant from sibilant fricatives, with the only exception being the similar values obtained for /s/ and voiceless non-sibilants /f, θ/. Although spectral peak location distinguished between post-alveolar /ʃ/ and alveolar fricatives /s, z/, it failed to distinguish among non-sibilants [12].

Tahiry pointed out that, the use of normalized energy in the frequency bands as an acoustic index for the classification of consonants, presents a better identification of Arabic stop consonants. Tahiry's results yielded to an overall classification of 90.27% stop consonants and more than 90% CV syllable recognition for all stops [13].

Our objective is to propose and study the performance of a classification system of Arabic fricative consonants; more precisely the consonants (/ف, f /, /خ, kh /, /ح, h /and / ش, sh /); using acoustic cues dependent on the energy distribution in the Syllables CV. We used Arabic fricatives because the articulatory space of fricatives in Arabic spans across most of the places of articulation in the vocal tract, starting from the lips and ending at the glottis.

The organization of this paper is as follows: The first section presents the methods and tools employed and the experiments carried out. The second section comprises the results found. In the last section, there is a summary of the findings and presentation of conclusions.

2   Methods

2.1   General Processing

We began with the construction of our corpus of Arabic language fricative consonants (/ف, f /, /خ, X, kh /, /ح, h /and /ش, ʃ, sh /). The fricatives were followed by each of three vowels /a, u, i/. Three Moroccan speakers (from the high school of technology) were asked to pronounce the sequences CVCVCV (C is an Arabic fricative consonant and V a vowel within /a/, /i/ or /u/) (see Fig. 1). Each CVCVCV token was repeated four times, yielding a total of 144 sequences (4 fricatives 3 vowels 3 speakers 4 repetitions). We then compute the energy band parameters in order to isolate each fricative consonant (see Fig. 2).

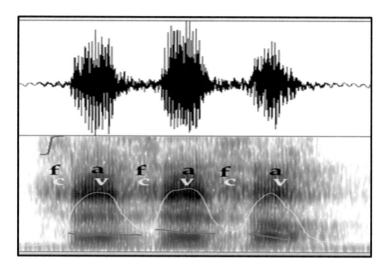

Fig. 1. Recording of fricative /f / with vowel /a/ using the Praat software

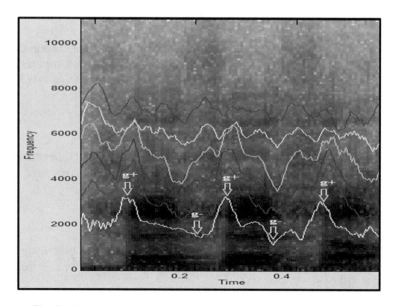

Fig. 2. Fricative consonants localized between g+ and g− landmarks.

2.2   Computation of Energy Band Parameters

Speakers were recorded in a sound proof booth with a high-quality microphone (Labtec AM-232). To prevent turbulence due to direct airflow from impinging on the microphone, the microphone was placed at approximately a 45° angle and 20 cm away from the corner of the speaker's mouth. All recordings sampled at 22050 Hz were divided into time segments of 11.6 ms with an overlap of 9.6 ms. 512-point fast Fourier transform (FFT) were then computed. The magnitude spectrum for each frame is smoothed by a 20-point moving average taken along the time index n. From the smoothed spectrum X(n, k), peaks in six different frequency bands (Band 1: 100–400 Hz; Band 2: 800–1500 Hz; Band 3: 1200–2000 Hz; Band 4: 2000–3500 Hz; Band 5: 3500–5000 Hz and Band 6: 5000–11025 Hz) were selected as :

$$E_b(n) = 10 \log 10 \, (max \, X \, | \, (n, k) \, |^2) \tag{1}$$

Where the band index b ranges from 1 to 6. The frequency index k ranges from the DFT indices representing the lower and upper boundaries for each band.

2.3   Computation of Rate of Change

Landmark detection involves measurement of rate of variation of a set of parameters extracted from the speech signal on a short-time basis, and locating regions with a significant variation characterizing the landmark. A rate of change measure based on first difference operation with a fixed time step is generally used to get the rate of variation of parameters. For band energy parameter Eb(n), ROC measure is defined as:

$$r_{Eb}(n) = E_b(n) - E_b(n - K) \tag{2}$$

Where K is the time step. This measure indicates the difference in parameter value of the current frame, from a frame preceding it by K frames. An abrupt transition is indicated by a well-defined peak in the ROC track, while the track has a very low value during steady-state segments.

2.4   Detection of Burst Onset Landmarks

The detection of voicing offsets (g−) and voicing onsets (g+) are performed using the peak energy variation in the frequency band from 0 to 400 Hz. The peak energy is computed as:

$$E_g(n) = 10 \log_{10}(max \, |X(n, k)|) \tag{3}$$

Where $k1 \leq k \leq k2$, k1 and k2 being the DFT indices corresponding to 0 and 400 Hz respectively. A rate of rise measure of $E_g(n)$ is computed with a time step of 50 ms (K = 50) as :

$$r_{Eg}(n) = E_g(n) - E_g(n - K) \tag{4}$$

The crossing points r_{Eg} (n) below and above threshold values of -9 dB and $+9$ dB respectively are taken as the voicing offset and voicing onset points. An intervocalic burst onset is located at the most prominent peak in the ROC, between the g$-$ and g+ points.

2.5   Computation of Normalized Energy Band

For each frame of the stop consonant, the normalized energy band is defined as:

$$E_{bn}(n) = E_b(n)/E_{Tn} \tag{5}$$

Where E_{bn} (n) is the normalized band energy b in the frame n, E(n) is the overall energy in the frame n and E_b (n) is the band energy b in the frame n.

3   Results and Discussions

3.1   Direct Acoustic Analysis

3.1.1   Energy Analysis of All Consonants in All Bands
We first showed the percentage of energy in the six bands (B1 [100 Hz–400 Hz], B2 [400 Hz–800 Hz], B3 [800 Hz–1500 Hz], B4 [1200 Hz–2000 Hz]) for the four Fricatives consonants (/ف, f /, /خ, kh /, /ح, h /and /ش, sh /). The results shows a strong energy presence in the bands (B5 and B6) for the consonant /ش, sh /. This energy tends towards zero for the other consonants in these bands (B5 and B6).

Thus, we note that the consonant /,ش, sh / is recognized by the presence of energy in the bands (B5 and B6). We then conducted other analyzes to classify the consonants (/ف, f /, /خ, kh /and /ح, h /).

3.1.2   Analysis of CV Syllables of Type (Xa, Xu, Xi)
In this analysis, we analyzed each consonant /ف, f/, /خ, kh /, /ح, h /and /ش, sh /with the three vowels / a, u, i /. We found that the energy in the middle of the vowel makes it possible to distinguish between the three vowels.

We have found that for a (fa / fu / fi) data set, the syllable that has zero energy in the middle of the vowel in the band (B2) will be classified as / fi /. The other syllables will be classified as / fu /or / fa /depending on the energy in the middle of the vowel in the band (B3). If this energy is zero, it is the syllable /fu /otherwise it is the syllable /fa / (see Fig. 3).

We have repeated the same analysis for the other consonants (sha /shu/shi), (kha / khu /khi) and (ha /hu /hi). We found that the energy distribution in the middle of the vowel is similar to that of the consonant (fa /fu /fi).

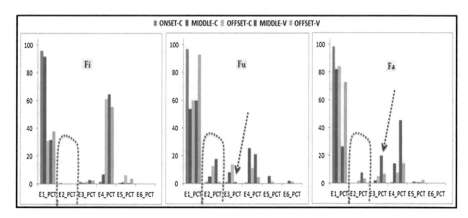

Fig. 3. Energy percentage in the bands for the consonant /f/ followed by the three vowels.

3.1.3   Analysis of Syllables CV of the Same Vowel

The analysis of the four fricative consonants (/ف, f /, /خ, kh /, /ح, h /and /ش, sh /) with the three vowels /a, u, i/ indicated that the consonant /ش, sh /is clearly recognized among the others consonants. On the other hand, for each of these four consonants X, the distribution of energy in the bands B2 and B3 makes it possible to distinguish between the syllables /Xa/, /Xu /and /Xi /.

To complete the classification of these fricative consonants, we have analyzed the energy distribution of the three consonants (/ف, f/, /خ, kh / and /ح, h /) accompanied by the same vowel.

/kha, fa, ha/

The analysis of the distribution of energy reveals that the syllable (/ kha /) is clearly distinguished from /ha /and /fa /. It is characterized by low energy in band B4 in the middle of the consonant.

The energy in the band B1 at the offset of the consonant makes it possible to discriminate between the syllables / fa /and / ha /. If the energy percentage at this position is less than or equal to 50%, the syllable is classified as / ha / if not, it is /fa /.

/khu, fu, hu/

The examination of the percentage of energy in the six bands for the syllables / khu, fu, hu /) indicates that they can be classified. If the percentage of energy in the band B3 in the middle of the vowel is greater than 10%, it is the syllable /hu /otherwise, it is the syllable / fu / or /khu /.

If the energy percentage in the B3 band at the end of the consonant (offset C) is zero, the syllable is / khu /otherwise it is / fu /.

/khi, fi, hi/

Considering the percentage of energy in the six bands for the syllables / khi, fi, hi / indicates a confusion in the classification of these three syllables. Indeed, no particularity was detected for these syllables.

Table 1. Results of decision acoustic analysis

Analysis set	Correctly classified instances	Confusion matrix					Result of the decision trees
All syllables	61.8056%	*a*	*b*	*c*	*d*	*classified as*	The consonant / sh, ش / is classified by the presence of energy in the bands B5 and B6
		49	35	5	19	*a = f*	
		32	55	4	17	*b = kh*	
		2	3	97	6	*c = sh*	
		22	18	2	66	*d = h*	
fa, fu, fi	93.5185%	*a*	*b*	*c*		*classified as*	The syllable / Xi / is classified by the percentage of energy in the middle of the vowel in the B2 band, and the syllables / Xu / and / Xa /are classified by the middle of the vowel in the B3 band.
		35	0	1		*a = fa*	
		1	35	0		*b = fi*	
		4	1	31		*c = fu*	
kha, khu, khi	91.6667%	*a*	*b*	*c*		*classified as*	
		33	2	1		*a = Kha*	
		2	33	1		*b = Khi*	
		4	1	33		*c = Khu*	X = (f, kh, sh ou h)
sha, shu, shi	82.4074%	*a*	*b*	*c*		*classified as*	
		31	1	4		*a = sha*	
		1	30	5		*b = shi*	
		5	3	28		*c = shu*	
ha, hu, hi	81.4815%	*a*	*b*	*c*		*classified as*	
		25	2	9		*a = ha*	
		2	34	5		*b = hi*	
		6	1	29		*c = hu*	
Kha, Fa, Ha	60.1852%	*a*	*b*	*c*		*classifiedas*	The syllable / Kha / is classified by the percentage of energy in the band B4 in the middle of the consonant (middle-C4 < 17%). If the energy at the middle-C4 > 17% the syllables / Fa / and / Ha /will be classified by energy in the B1 band at offset of the consonant (offset-C1 < 49% for / Ha / the opposite for / Fa /)
		17	11	8		*a = fa*	
		7	22	7		*b = kha*	
		2	8	26		*c = ha*	
Khu, Fu, Hu	76.8519%	*a*	*b*	*c*		*classified as*	The syllable / Hu / is classified by the percentage of energy in the band B3 in the middle of the vowel (medium-V3 > 9%). If the energy at the medium-V3 < 9%, the syllables / Fu / and / Khu / will be classified by energy in the B3 band at the offset of the consonant (offset-C3 < 3% for / Khu /the opposite for / Fu/)
		26	6	4		*a = fu*	
		7	27	2		*b = khu*	
		3	3	30		*c = hu*	
Khi, Fi, Hi	59.2593%	*a*	*b*	*c*		*classified as*	The decision tree showed a confusion between the syllables
		17	7	12		*a = fi*	
		5	26	5		*b = khi*	
		12	3	21		*c = hi*	

3.2   Decision Acoustic Analysis

To validate and measure the performances of our classifier, we have taken the same approach using the algorithm j48 (decision tree) of the Weka software.

The application of the algorithm j48 confirms the results obtained in the acoustic analysis with a very interesting recognition rate (Table 1).

The previous studies of the fricative consonants allow a general classification of these consonants (voiced/unvoiced) or (sibilant/non-sibilant), on the other hand our algorithm allows a clear classification between the Arabic fricatives consonants. This algorithm is able to identify one of the four consonants (/ف, f /, /خ, kh /, /ح, h /and /ش, sh /) followed by the three vowels.

4   Conclusion

The Direct acoustic analysis allows to develop a classifier capable of recognizing all syllables except / fi /, / khi /and / hi /. Decision analysis using, algorithm j48 of Weka software confirmed the results of the direct acoustic analysis. The elaborate classifier has been validated with interesting performance rates for all syllables. For the consonant / Ch., ش /, the recognition rate is close to 90% (97 out of 108). For the other three consonants (/ف, f /, /خ, kh /and /ح, h /), this classifier allows to recognize the vowel that accompanies the consonant with a recognition rate higher than 82%. It is also possible to classify the syllables (/khu/, /Hu/and /Fu/) with a recognition rate of 77%, the syllables (/kha/, /ha/and /fa/) with a recognition rate of 60% and syllables (/ shi/, /hi/ and /fi/) with a recognition rate of 59%.

References

1. Stevens, K.N.: Airflow and turbulence noise for fricative and stop consonants. J. Acoust. Soc. Am. 50, 1180–1192 (1971)
2. Shadle, C.: The acoustics of fricative consonants, in Research Laboratory of Electronics, Technical Report 506 MIT, Cambridge, MA (1985)
3. Strevens, P.: Spectra of fricative noise in human speech. Lang. Speech 3, 32–49 (1960)
4. Behrens, S.J., Blumstein, S.E.: Acoustic characteristics of English voiceless fricatives: a descriptive analysis. J. Phonet. 16, 295–298 (1988)
5. Nittrouer, S., Studdert-Kennedy, M., McGowan, R.S.: The emergence of phonetic segments: evidence from the spectral structure of fricative-vowel syllables spoken by children and adults. J. Speech Lang. Hear. Res. 32, 120–132 (1989)
6. Tjaden, K., Turner, G.S.: Spectral properties of fricatives in amyotrophic lateral sclerosis. J. Speech Lang. Hear. Res. 40, 1358–1372 (1997)
7. Jongman, A., Wayland, R., Wong, S.: Acoustic characteristics of English fricatives. J. Acoust. Soc. Am. 108, 1252–1263 (2000)
8. Nittrouer, S.: Children learn separate aspects of speech production at different rates: evidence from spectral moments. J. Acoust. Soc. Am. 97, 520–530 (1995)
9. Tomiak, G.R.: An acoustic and perceptual analysis of the spectral moments invariant with voiceless fricative obstruents, Doctoral dissertation, SUNY Buffalo (1990)

10. Shadle, C.H., Mair, S.J.: Quantifying spectral characteristics of fricatives. In: Proceedings of the International Conference on Spoken Language Proceedings (ICSLP), pp. 1521–1524 (1996)
11. Elina, N.: Acoustic characteristics of Greek fricatives. J. Acoust. Soc. Am. **135**, 2964–2976 (2014)
12. Al-khairy, A.M.: Acoustic characteristics of Arabic fricatives. University of Florida (2005)
13. Tahiry, K.: Arabic stop consonants characterization and classification using the normalized energy in frequency bands. Int. J. Speech Technol. **20**, 869–880 (2017)

Cancer Classification Using Gene Expression Profiling: Application of the Filter Approach with the Clustering Algorithm

Sara Haddou Bouazza[1]([✉]), Khalid Auhmani[2], Abdelouhab Zeroual[1], and Nezha Hamdi[1]

[1] Department of Physics, Faculty of Sciences Semlalia, Cadi Ayyad University, Marrakech, Morocco
Sara.hb.sara@gmail.com, zeroual@uca.ma,
nezha_hamdi@yahoo.com
[2] Department of Industrial Engineering, National School of Applied Sciences, Cadi Ayyad University, Safi, Morocco
kauhmani@yahoo.com

Abstract. In this paper, we investigate the classification accuracy of different cancers based on microarray expression values. For this purpose, we have used hybridization between a filter selection method and a clustering method to select relevant features in each cancer dataset. Our work is carried out in two steps. First, we examine the effect of the filter selection methods on the classification accuracy before clustering. The studied filter selection methods are SNR, ReliefF, Correlation Coefficient and Mutual Information. The K Nearest Neighbor, Support Vector Machine and Linear Discriminant Analyses classifier were used for supervised classification task.

In the second step, the same investigation is carried out, but the feature selection task is preceded by a k-means clustering operation.

Obtained results showed that the best classification accuracies were obtained (for leukemia, colon, prostate, lung and lymphoma cancers datasets) for SNR method. After adding the clustering step to the phase of the feature subset selection, the classification accuracy has been increased for the four selection methods SNR, ReliefF, Correlation Coefficient, and Mutual Information.

Keywords: DNA microarray · Feature selection · Supervised classification
Clustering · Image processing

1 Background

DNA microarrays are characterized by high dimensionality due to the high number of features composed of thousands of genes and a limited number of observations. For this reason, it becomes necessary to reduce the dimensionality of dataset in order to decrease the size of the dataset matrix and also, to make the classification task easier and faster.

One form of the dimensionality reduction is feature subset selection, an imperative step in the field of classification.

So, in order to classify a cancer dataset, we need to select the relevant features that best represent the cancer dataset. To do this, we need to use a filter selection method on

© Springer International Publishing AG, part of Springer Nature 2018
A. Abraham et al. (Eds.): SoCPaR 2017, AISC 737, pp. 90–99, 2018.
https://doi.org/10.1007/978-3-319-76357-6_9

the original cancer dataset, and then use this selected subset of features to classify cancer dataset. The classification accuracy obtained by the classifier represents the performance of the subset selected.

In this paper, we suggest to use the k means clustering not only as a classification algorithm but also as a selection method. We hybridized between a filter selection method (the signal to noise ratio (SNR), ReliefF, Correlation Coefficient (CC), ReliefF and Mutual Information (MI)) and the clustering K-means. To compare these feature selection methods, an evaluation of the dimensionality reduction had been done using four supervised classifiers (k nearest neighbors (KNN), Support Vector Machine (SVM) and Linear Discriminant analysis (LDA)).

The goal of this hybridization is to improve classification performance and to accelerate the search to identify important feature subsets.

2 Related Works

Features selection methods become the focus of much research in areas of application for which datasets with thousands of features are available. Some of the used methods in the field of feature selection are:

- Fisher, T-statistics, Signal to noise ratio and ReliefF selection methods [1].
- The use of two-step neural network classifier [2].
- The (BW) discriminant score was proposed by [3]. It is based on the dispersion ratio between classes and intra-class dispersion.
- A hybridization between Genetic Algorithm (GA) and Max-relevance, Min-Redundancy (MRMR) [4].

3 Materials and Methods

We used different feature selection methods and classifiers for cancer classification.

In the first step, we downloaded the dataset of each cancer composed of thousands of features. In the second step we reduced the number of features, using a feature subset selection, to only relevant features. In the final step, we classify the datasets.

3.1 Dataset Description

In this paper, we studied the effect of feature selection methods on three commonly used gene expression datasets: leukemia cancer, Colon cancer and Prostate cancer (Table 1).

- Leukemia is composed of 7129 features and 72 samples. It contains two classes: acute lymphocytic leukemia (ALL) and acute myelogenous leukemia (AML). It can be downloaded from the website[1].

---

[1] broadinstitute.org/cgi-bin/cancer/publications/pub_paper.cgi?mode = view&paper_id = 43.

Table 1. Datasets and parameters used for experiments

Dataset	No. of features	No. of observation	No. of classes
Leukemia [5]	7129	72	2
Colon [6]	6500	62	2
Prostate [7]	12600	101	2
Lung [8]	12533	181	2
Lymphoma [9]	7070	77	2

- Colon cancer is composed of 6500 features and 62 samples. It contains two classes: Tumor and Not tumor. It can be downloaded from this website[2].
- Prostate cancer is composed of 12600 features and 101 samples. It contains two classes: Tumor and Not tumor. It can be downloaded from this website[3].
- Lung Cancer is composed of 12533 features and 181 samples; it contains two classes: malignant pleural mesothelioma (MPM) and adenocarcinoma (ADCA). Data could be downloaded from the website[4].
- Lymphoma cancer is composed of 7070 genes and 77 samples. It contains two classes: diffuse large B-cell lymphoma (DLBCL) and follicular lymphoma (FL). It is available to the public at the website[5].

3.2   Feature Subset Selection

Feature selection is the process of selecting a subset of relevant features for model construction (Fig. 1).

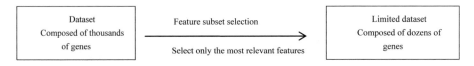

Fig. 1. Feature subset selection

The main idea to apply a feature selection method is that the dataset contains many Features that are either redundant or irrelevant and can consequently be removed without high loss of information.

A feature selection algorithm can be considered as the combination of a search technique for proposing new feature subsets, along with an evaluation measure which scores the different feature subsets. There are three main categories of feature selection algorithms: wrappers, filters and embedded methods [10].

---

[2] genomics-pubs.princeton.edu/oncology/affydata/insdex.html.

[3] broadinstitute.org/cgi-bin/cancer/publications/pub_paper.cgi?mode = view&paper_id = 75.

[4] http://www.chestsurg.org.

[5] http://www.broadinstitute.org/cgi-bin/cancer/datasets.cgi.

- Wrapper methods use a predictive model to score feature subsets.
- Filter methods use a proxy measure instead of the error rate to score a feature subset.
- Embedded methods are a catchall group of techniques which perform feature selection as part of the model construction process.

We are interested in this paper with filter methods which are based on the estimated weight (scores) corresponding to each feature (gene) used to order then to select the most relevant descriptors.

The methods used in this work are the Signal to Noise Ratio (SNR), Correlation Coefficient (CC), ReliefF, Mutual Information (MI) and clustering (K-means).

The signal to noise ratio

The signal to noise ratio, called also S/R test, recognizes relevant features by calculating the score S/R of each gene (g) [11].

This score was proposed by [5] and expressed as follows:

$$S/R_{(g)} = \frac{M_{1g} - M_{2g}}{S_{1g} + S_{2g}} \tag{1}$$

Where M_{kg} and S_{kg} denote the mean and the standard deviation of the feature g for samples of classes 1 and 2.

ReliefF

This algorithm presented as Relief [12] and then developed and adjusted to the multi-class case by Kononenko as the ReliefF [13].

This criterion measures the ability of each feature to group data of the same class and discriminating those having different classes. The algorithm is described as follows:

- Initialize the score (or the Weight) wd = 0, d = 1,..., D
- For t = 1 ...N
- Pick randomly an instance x_i
- Find the k nearest neighbors to x_i having the same class (hits)
- Find the k nearest neighbors to x_i having different class (misses c)
- For each feature d, update the weight:

$$W_d = wd - \sum_{j=1}^{K} \frac{\text{diff}(x_i, d, \text{hits}_j)}{m * k} + \sum_{c \neq \text{class}(x_i)} \frac{p(c)}{1 - p(\text{class}(x_i))} \sum_{j=1}^{k} \frac{\text{diff}(x_i, d, \text{misses}_j)}{m * k} \tag{2}$$

The distance used is defined by:

$$\text{diff}(x_i, d, x_j) = \frac{|x_{id} - x_{jd}|}{\max(d) - \min(d)} \tag{3}$$

Max (d) (resp. min (d)) is the maximum (resp. minimum) value that may take the feature designated by the index d on the data set. x_{id} is the value of the d_{th} feature of the data x_i.

This method does not eliminate redundancy, but defines a relevant criterion.

Correlation Coefficient
Correlation coefficients measure the strength of association between two features. The Pearson correlation coefficient measures the strength of the linear association between features [14].

Let and S_y be the standard deviations of two random features X and Y respectively. Then the Pearson's product moment correlation coefficient between the features is:

$$\rho_{x,y} = \frac{cov(X, Y)}{S_x S_y} = \frac{E((X - E(X))(Y - E(Y)))}{S_x S_y} \tag{4}$$

Where cov(.) means covariance and E(.) denotes the expected value of the feature.

Mutual Information
Let us consider a random feature G that can take n values over several measures, we can empirically estimate the probabilities $P(G_1)$, ..., $P(G_n)$ for each state G_1, ..., Gn of feature Shannon's entropy [15] of the feature is defined as:

$$H_{(G)} = -\sum_{i=0}^{NG} P_{(G)} \log (P_{G(i)}) \tag{5}$$

The mutual information measures the dependence between two features. In the situation of genes selection, we use this measure to recognize genes which are related to the class C. The mutual information between C and one gene G is measured by the following expression:

$$MI(G, C) = H(G) + H(C) - H(G, C) \tag{6}$$

$$H(G, C) = -\sum_{i=0}^{NG} - \sum_{j=0}^{NG} P_w(i, j) \log(P_w(i, j)) \tag{7}$$

Cluster analysis
Cluster analysis or clustering is the task of assembling a set of objects in such a way that objects in the same group are more similar to each other than to those in other groups. In clustering, the k-means algorithm can be used to divide the input data set into k groups or clusters and returns the index of the cluster to which it has assigned each feature.

K-means algorithm is described as follows:

Given an initial set of k means $m_1(1),...,m_k(1)$, the algorithm proceeds by alternating between two steps [16]:

- Assignment step: Assign each feature to the cluster whose mean yields the least within-cluster sum of squares. Since the sum of squares is the squared Euclidean distance, this is intuitively the "nearest" mean

$$S_i^{(t)} = \left\{ x_p : \left\| x_p - m_i^{(t)} \right\|^2 \leq \left\| x_p - m_j^{(t)} \right\|^2, \ 1 \leq j \leq k \right\} \tag{8}$$

Where each x_p is assigned to exactly one $S^{(t)}$, even if it could be assigned to two or more of them.

- Update step: Calculate the new means to be the centroids of the features in the new clusters.

$$m_i^{(t+1)} = \frac{1}{|S_i^t|} \sum_{x_j \in S_i^t} x_j \tag{9}$$

3.3 Classification

To compare all feature selection methods, an evaluation of the dimensionality reduction was done using a supervised classification of the three cancers.

Supervised classification is the process of discriminating data, a set of objects or data more widely, so that the objects in the same class are closer to each other than other classes.

To study the performances of the selected features methods, we used the KNN (K nearest neighbors) classifier.

K Nearest Neighbors

K nearest neighbors' is a classifier that stores training samples and classifies the test samples based on a similarity measure.

In K Nearest Neighbors, we try to find the most similar K number of samples as nearest neighbors in a given sample, and predict class of the sample according to the information of the selected neighbors.

We can compute the Euclidean distance between two samples by using a distance function $D_E(X, Y)$, where X, Y are samples composed of N features, such that $X = \{X_1, \ldots, X_N\}, Y = \{Y_1, \ldots, Y_N\}$.

$$D_E(X,Y) = \sum_{j=1}^{k} \sqrt{(X_i^2 - Y_i^2)} \tag{10}$$

Support Vector Machines (SVM)

Support vector machines are supervised learning models used for supervised classification [17]. Support Vector Machines are based on two key concepts: the notion of maximum margin and the concept of kernel functions.

Linear Discriminant Analysis (LDA)

Linear Discriminant Analysis is an algorithm used in machine learning to search and find a linear combination of features that characterizes or separates two or more classes of objects [18].

To evaluate the performances of the classifiers, we measure the value of the classification accuracy $A_{ccuracy}$ [19]:

$$A_{ccuracy} = 100 * (TP + TN)/(TN + TP + FN + FP) \tag{11}$$

Where TP is true positive for correct prediction to disease class, TN is true negative for correct prediction to normal class, FP is false positive for incorrect prediction to disease class, and FN is false negative for incorrect prediction to normal class.

All the algorithms used in this paper have been run using (MATLAB)

4   Results

In this section, we report the results of an experimental study of the effect of the k-means clustering on five commonly used gene expression datasets.

Each dataset is characterized by a group of features, those features are the genes.

After dividing the initial dataset into training data and test data, we applied a subset selection method on training data to select the most relevant features. This subset helps to classify dataset using a classifier (KNN, SVM and LDA). Test data is used to investigate the performances of selection methods and classifiers.

To increase the selection methods performances, we add a clusterisation to the selection step. We divide training data into clusters, and then we select relevant features in each cluster. The obtained subset presents the most relevant features in the dataset.

Tables 2, 3, 4, 5 and 6 compares the classification accuracy obtained (for leukemia, colon, prostate, lung and lymphoma cancers, respectively) before and after adding the k-means clustering to the selection step.

We can clearly remark the advantage of adding the clusterisation step to the feature selection process. It increases the accuracy of the four selection methods investigated in this paper.

Table 2. Performance of comparison for proposed classifiers (leukemia cancer)

Classifier	KNN				SVM				LDA			
Selection method	Before clustering		After clustering		Before clustering		After clustering		Before clustering		After clustering	
	Max accuracy (%)	Features selected	Max accuracy (%)	Features selected	Max accuracy (%)	Features selected	Max accuracy (%)	Features selected	Max accuracy (%)	Features selected	Max accuracy (%)	Features selected
SNR	100	13	100	5	97.05	4	100	4	100	9	100	5
ReliefF	100	41	100	8	97.05	2	100	3	100	69	100	21
CC	100	50	100	19	97.05	2	97.05	2	100	93	100	35
MI	76.41	56	91.1	18	84.2	5	91.1	5	91.1	10	94.1	5

Table 3. Performance of comparison for proposed classifiers (colon cancer)

Classifier	KNN				SVM				LDA			
Selection method	Before clustering		After clustering		Before clustering		After clustering		Before clustering		After clustering	
	Max accuracy (%)	Features selected	Max accuracy (%)	Features selected	Max accuracy (%)	Features selected	Max accuracy (%)	Features selected	Max accuracy (%)	Features selected	Max accuracy (%)	Features selected
SNR	92.8	5	**95**	6	85.7	29	**100**	4	92.8	2	**100**	8
ReliefF	85.7	40	**95**	25	85.7	11	**92.8**	7	78.5	78	**92.8**	15
CC	92.8	7	**94.2**	2	85.7	2	**95**	2	92.8	27	**95**	14
MI	85.7	43	**95**	25	78.5	5	**91.1**	5	71.4	19	**94.1**	3

Table 4. Performance of comparison for proposed classifiers (prostate cancer)

Classifier	KNN				SVM				LDA			
Selection method	Before clustering		After clustering		Before clustering		After clustering		Before clustering		After clustering	
	Max accuracy (%)	Features selected	Max accuracy (%)	Features selected	Max accuracy (%)	Features selected	Max accuracy (%)	Features selected	Max accuracy (%)	Features selected	Max accuracy (%)	Features selected
SNR	90	22	**90**	1	92	8	**100**	9	100	4	**100**	3
ReliefF	90	32	**90**	5	92	34	**92**	7	100	75	**100**	43
CC	85	6	**90**	1	92	44	**92**	5	100	6	**100**	3
MI	65	1	**90**	4	58.8	56	**78.4**	10	92	10	**95**	8

Table 5. Performance of comparison for proposed classifiers (lung cancer)

Classifier	KNN				SVM				LDA			
Selection method	Before clustering		After clustering		Before clustering		After clustering		Before clustering		After clustering	
	Max accuracy (%)	Features selected	Max accuracy (%)	Features selected	Max accuracy (%)	Features selected	Max accuracy (%)	Features selected	Max accuracy (%)	Features selected	Max accuracy (%)	Features selected
SNR	100	6	**100**	3	100	33	**100**	10	100	64	**100**	14
ReliefF	100	21	**100**	4	100	17	**100**	11	99.3	80	**100**	28
CC	100	28	**100**	5	100	36	**100**	12	100	82	**100**	19
MI	83.2	10	**96.6**	9	88.5	5	**90.6**	5	96.6	24	**99.3**	20

Table 6. Performance of comparison for proposed classifiers (lymphoma cancer)

Classifier	KNN				SVM				LDA			
Selection method	Before clustering		After clustering		Before clustering		After clustering		Before clustering		After clustering	
	Max accuracy (%)	Features selected	Max accuracy (%)	Features selected	Max accuracy (%)	Features selected	Max accuracy (%)	Features selected	Max accuracy (%)	Features selected	Max accuracy (%)	Features selected
SNR	100	4	**100**	3	100	32	**100**	10	100	24	**100**	12
ReliefF	100	86	**100**	12	100	2	**100**	1	100	93	**100**	17
CC	100	13	**100**	8	100	39	**100**	4	100	97	**100**	22
MI	86.9	10	**95.6**	7	86.9	15	**97**	7	52.1	50	**99.3**	4

From the results obtained in Tables 2, 3, 4, 5 and 6 we remark that the k means clustering step obtains a substantial reduction in feature set size maintaining better accuracy compared with results before clustering step, for the chosen Gene datasets of high dimensionality.

5   Conclusion and Discussion

We have shown in this paper that feature selection methods can be applied successfully to a classification situation, using only a limited number of training samples in a high dimensional space of thousands of features.

We performed several studies on leukemia, colon, prostate, lung and lymphoma cancer datasets. The objective was to classify datasets of each cancer into two classes.

The experimental results show that the proposed method has efficient searching strategies and is capable of producing a good classification accuracy with a small and limited number of features simultaneously.

The best result obtained for leukemia cancer is an accuracy of 100% for only 5 genes. For Colon cancer, we obtain 95% for only 6 genes. For prostate cancer, we obtain 90% for 1 gene. For lung cancer, we obtain 100% for 3 genes. For lymphoma cancer, we obtain 100% for 3 genes.

These results encourage adding a clusterisation before the selection step. It increases the classification accuracies and decreases the number of features selected.

References

1. Bouazza, S.H., Hamdi, N., Zeroual, A., Auhmani, K.: Gene-expression-based cancer classification through feature selection with KNN and SVM classifiers. In: 2015 Intelligent Systems and Computer Vision (ISCV) (2015)
2. Vincent, I., Kwon, K.-R., Lee, S.-H., Moon, K.-S.: Acute lymphoid leukemia classification using two-step neural network classifier, May 2015
3. Logique floue et algorithmes génétiques pour le pré-traitement de données de biopuces et la sélection de gènes, thèse de doctorat, edmundobonilla huerta (2008)
4. El Akadi, A.: Contribution to select relevant features in supervised classification: application to the selection of genes for DNA chips and facial characteristics (2012)
5. Zhang, L., Chen, Y., Abraham, A.: Hybrid flexible neural tree approach for leukemia cancer classification. In: World Congress on Information and Communication Technologies (2011)
6. Park, C., Cho, S.B.: Evolutionary ensemble classifier for lymphoma and colon cancer classification. In: Conference: Evolutionary Computation (2003). https://doi.org/10.1109/CEC.2003.1299385
7. Singh, D., Febbo, P.G., Ross, K., Jackson, D.G., Manola, J., Ladd, C., Tamayo, P., Renshaw, A.A., D'Amico, A.V., Richie, J.P., Lander, E.S., Loda, M., Kantoff, P.W., Golub, T.R., Sellers, W.R.: Cancer Cell: March 2002, vol. 1, 28 Feb 2002
8. Gordon, G.J., Jensen, R.V., Hsiao, L.L., Gullans, S.R., Blumenstock, J.E., Ramaswamy, S., Richards, W.G., Sugarbaker, D.J., Bueno, R.: Translation of microarray data into clinically relevant cancer diagnostic tests using gene expression ratios in lung cancer and mesothelioma. Cancer Res. **62**, 4963–4967 (2002)
9. Shipp, M.A., Ross, K.N., Tamayo, P., et al.: Diffuse large B-cell lymphoma outcome prediction by gene-expression profiling and supervised machine learning. Nat. Med. **8**(1), 68–74 (2002)
10. Guyon, I., Elisseeff, A.: An introduction to variable and feature selection. JMLR **3**, 1157–1182 (2003)
11. Cuperlovic-Cuf, M., Belacel, N., Ouellette, R.J.: Determination of tumour marker genes from gene expression data. DDT **10**(6), 429–437 (2005)

12. Kira, K., Rendell, L.: A practical approach to feature selection, pp. 249–256 (1992)
13. Robnik-Šikonja, M., Kononenko, I.: Theoretical and empirical analysis of relieff and rrelieff. Mach. Learn. **53**(1–2), 23–69 (2003)
14. Egghe, L., Leydesdorff, L.: The relation between Pearson's correlation coefficient r and Salton's cosine measure. J. Am. Soc. Inf. Sci. Technol. **60**, 1027–1036 (2009). https://doi.org/10.1002/asi.21009
15. Shannon, E.: A mathematical theory of communication. Bell Syst. Tech. J. **27**, 623–654 (1948)
16. MacKay, D.: An example inference task: clustering. Information Theory, Inference and Learning Algorithm, pp. 284–292. Cambridge University Press, Cambridge (2003). Chapter 20. ISBN 0-521-64298-1. MR 2012999
17. Smola, A.J., Schölkopf, B.: A tutorial on support vector regression. Stat. Comput. **14**(3), 199–222 (2004)
18. Sergey, Y.: Sensors and biosensors, MEMS technologies and its applications. In: Advances in Sensors: Reviews, vol. 2. Par Sergey Yurish (2014)
19. Pehlivanlı, A.Ç.: A novel feature selection scheme for high-dimensional data sets: four-staged feature selection. J. Appl. Stat. **43**, 1140–1154 (2015)

Detection of Negative Emotion Using Acoustic Cues and Machine Learning Algorithms in Moroccan Dialect

Abdellah Agrima[1(✉)], Laila Elmazouzi[2], Ilham Mounir[2],
and Abdelmajid Farchi[1]

[1] IMII Laboratory, Faculty of Sciences and Technics, University Hassan First,
Settat, Morocco
agrima.abdellah@gmail.com,
abdelmajid.farchil@gmail.com
[2] LAPSSII Laboratory, Graduate School of Technology,
University Cadi Ayyad, Safi, Morocco
elmazouzi2001@yahoo.fr, ilhamounir@gmail.com

Abstract. The speech signal provides rich information about the speaker's emotional state. Therefore, recognition of the emotion in speech has become one of the research themes in the processing of speech and applications based on human-computer interaction. This article provides an experimental study and examines the detection of negative emotions such as fear and anger with regard to the neutral emotional state. The data set is collected from speeches recorded in the Moroccan Arabic dialect. Our aim is first to study the effects of emotion on the selected acoustic characteristics, namely the first four formants F1, F2, F3, F4, the fundamental frequency F0, Intensity, Number of pulses, Jitter and Shimmer and then compare our results to previous works. We also study the influence of phonemes and speaker gender on the relevance of these characteristics in the detection of emotion. To this aim, we performed classification tests using the WEKA software. We found that F0, Intensity, Number of Pulses have the best rates of recognition regardless speaker gender and phonemes. Moreover the second and third formant are the features that highlighted phoneme's effect.

Keywords: Emotion · Classification · Speech processing
Fundamental frequency · Formant · Intensity
Number of pulses · Jitter · Shimmer

1 Introduction

Emotion has an impact on our judgment (for example [10]) and our reasoning (for example [12]). It also affects attention, motivation, and memory, problem-solving or decision-making. In recent years, more and more researchers are interested in the study of emotions in the voice (for example [27]) with often the same way of proceeding: using data labeled in emotion, a set of indices is checked and the methods Data search are used to recognize emotions. In the same framework, using the Moroccan dialect, we

© Springer International Publishing AG, part of Springer Nature 2018
A. Abraham et al. (Eds.): SoCPaR 2017, AISC 737, pp. 100–110, 2018.
https://doi.org/10.1007/978-3-319-76357-6_10

want to examine the relevance of certain voice signal indices such as the first four formants F1, F2, F3 and F4, the fundamental frequency F0, Intensity, Number of pulses, Jitter and Shimmer in detecting negative emotions namely anger and fear related to the neutral state. We also study the effect of phonemes and gender of speaker.

In the literature (for example [11, 19, 20, 23, 29] etc.) similar results have been reported concerning the acoustic indices that characterize different emotional states. The main correlates presented in these articles related to this study are:

- For anger, the mean of the fundamental frequency F0 is increased (for example [23, 25, 28]). F1 is raised (for example [31]). Also Banse and Scherer (for example [3]) found that angry speech is shown through heightened energy values in the higher frequencies.
- Fear is difficult to detect and identify (for example [3]). It is characterized by the strong increase in the mean of the fundamental frequency (F0) (for example [3, 28]). However, compared with anger, the average F0 is lower (for example [18]). F1 is raised (for example [28]). Jitter is activated (for example [6]).
- Some voice cues like Jitter are more difficult to simulate for an actor than others (for example [20]).

Currently, several studies use emotional classification systems. These systems are based on learning methods because of their ability to learn, from a sufficient amount of acoustic data, the properties of each state of emotion (for example [1, 2, 4, 9, 13, 16–18, 21, 22, 24, 26, 30, 32, 33].

In the same context and in order to further verify our hypotheses and observations on the studied parameters, we used three learning algorithms to elaborate classification models.

2   Methodology: Participants and Materials and Procedure

In order to verify the effect of emotions on acoustic indices, 200 speakers (100 women, 100 men) were invited to imitate negative emotions, namely fear and anger as well as neutral state, pronouncing repeatedly two sentences in the Moroccan dialect. The group included participants between the ages of 20 and 22 years, all of whom are from the central region of Morocco and had at least 12 years of education.

Since we wanted to examine the effect of emotion on phonemes, we chose two sentences containing two generic words from the central Moroccan dialect: "Safi" (which means "enough") and "Maymkench" (which means "impossible"). The first word contains the fricatives "s" and "f" and the second contains the plosive consonant "k". It should be noted that at this level we have not carried out an in-depth study on the nature of the phonemes nor on the fundamental unit of segmentation of the word since this not the main objective of this work. The recordings were examined from a perceptual point of view by people who had no idea about the experience. As a result, we selected 108 speakers (54 women and 54 men) and for each speaker we selected only 3 records for each sentence and each emotion.

We then proceeded to a manual segmentation of the acoustic signal in which we have isolated the two words "Safi" and "Maymkench". For all acoustic indices extracted, we calculated the mean obtained from the three records.

For the record, a mobile phone was used and maintained at a distance of 15 cm from the mouth. Experiments were carried out in the laboratory. Features were extracted from the recorded speech using Praat (for example [5]). Voice data was digitized at a sampling frequency of 8 kHz. For each utterance we extracted the first four formants (F1, F2, F3, and F4), the fundamental frequency F0, the intensity, the number of pulses, Jitter and Shimmer. To test the relevance of each characteristic in the determination of the emotions we used three classifiers: SVM (vector machine support), RN (artificial neural networks) and J48 (decision tree) using WEKA (for example [15]).

3   Results

a. **Analysis of acoustic features measurements**

Firstly, we studied the effect of emotional state on each feature and for each speaker. We present here a selection of features that shows high significance for the expression of negative emotions according to phonemes and speaker gender. Figures (1, 2, 3, 4, 5 and 6) summarize the results obtained for six subjects (three males, three females).

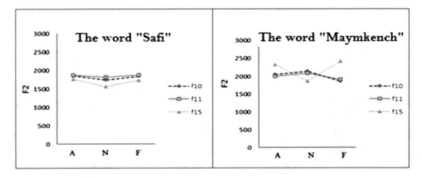

Fig. 1. Formant F2 for three speakers (female 10, female 11, female 15) for the words «Safi» and «Maymkench».

By analyzing the results we obtained for the words "Safi" and "Maymkench" for each feature, we noted that:

- The values of the first formant differ according to phonemes and gender speaker. For females, F1 corresponding to the word "Safi" is greater than that of the word "Maymkench". However, the same behavior is observed in both cases: F1Neutral < F1Fear < F1Anger. The same findings were reported in (for example [3, 23, 25, 28, 31]). For males, the results were less conclusive.

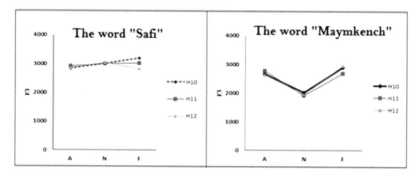

Fig. 2. Formant F3 for three speakers (male 10, male 11, male 12) for the words «Safi» and «Maymkench».

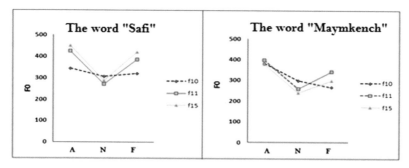

Fig. 3. F0 mean for three speakers (female 10, female 11, and female 15) for the words «Safi» and «Maymkench».

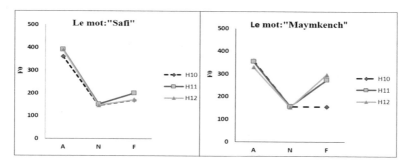

Fig. 4. F0 mean for three speakers (male 10, male 11, and male 12) for the words «Safi» and «Maymkench».

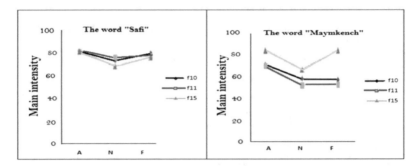

Fig. 5. Mean intensity for three speakers (female 10, female 11, and female 15) for the words «Safi» and «Maymkench».

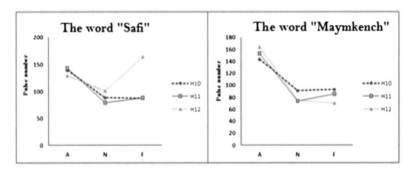

Fig. 6. Number of pulses for three speakers (male 10, male 11, male 12) for the words «Safi» and «Maymkench».

- For females, the behavior of the second formant depends closely on the phoneme and the emotion (for example Fig. 1). It is noteworthy that the F2 values of the word "Safi" vary in a reduced frequency band for all three emotions, which probably creates confusion between anger and fear. This band is larger for the word "Maymkench". For males, no significant changes were observed according to emotions and phonemes.
- The values of the third formant corresponding to males show a clear dependence on phonemes. Indeed, for the word "Maymkench", a general trend is observed F3Neutral < F3Anger < F3Fear. In the opposite, the values obtained for the word "Safi" are very close independently of emotion (for example Fig. (2)). For females, the values of F3 do not seem to be affected nor by the emotion neither by the phoneme.
- The mean of the fundamental frequency F0 maintains an identical behavior for the two words independently of speaker gender: F0Neutral < F0Fear < F0Anger. Moreover, the values corresponding to the word "Maymkench" are slightly lower than those of the word "Safi" (for example Figs. (3 and 4)).

- Again, independently of gender speaker, the values of the mean intensity (MI) indicate a clear dependence on phonemes and emotions. Contrary to the word "Safi", the results obtained for the word "Maymkench" allow to distinguish between the three emotions MI Neutral < MI Fear < MI Anger (for example Fig. (5)).
- The number of pulses (NP) observes almost the same behavior for the two words: NP Neutral < NP Fear < NP Anger. Moreover, the values vary in the same range (for example Fig. (6)).
- No general behavior was observed for Jitter and Shimmer. The values obtained for the two words vary in the same data range.

4 Classification Models

In order to verify the accuracy of these observations, we used the classification algorithms provided by WEKA on all the data collected for the corpus. We chose three algorithms, namely neural networks (Perceptron Multilayer (RN)), Support Vector Machines (SMO) and Decision Trees (J48). For each feature, tests were first performed with 10-folds cross-validation. In the case of intensity and fundamental frequency, we included minimal, maximal and mean values in the process of classification. Our approach consists of two steps. Firstly, we applied these algorithms on each characteristic in order to study its relevance in the detection of the emotions according to phonemes and speaker gender. Secondly, we used the same techniques on all the studied features to see the improvement they may bring to the quality of the detection of emotions.

a. **Specific classification models: 10-fold cross validation (CV)**

By analyzing Figs. 7 and 8, we observed that the best recognition rates are given by the characteristics F0, F2, F3, INT, and NP. However, these rates have a different ranking according to the type of phonemes and speaker gender (Table 1).

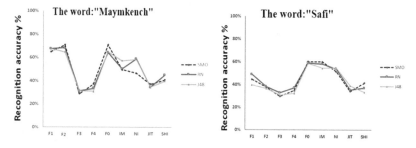

Fig. 7. Recognition accuracy for the three classifiers provided for the mean of the fundamental frequency F0 and the first formants F1, F2, F3 and F4, intensity (INT), number of pulses (NI), Jitter and Shimmer for the words: «safi» and «maymkench» (females).

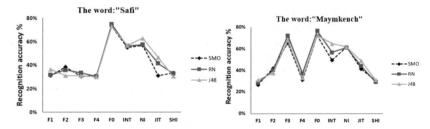

Fig. 8. Recognition accuracy for the three classifiers provided for the mean of the fundamental frequency F0 and the first formants F1, F2, F3 and F4, intensity (INT), number of pulses (NP), Jitter and Shimmer for the words: «safi» and «maymkench» males.

Table 1. Ranking of accuracies according to phonemes and speaker gender.

Speaker gender	Ranking	1st	2nd	3rd	4th
Females	"Safi"	F0	INT	NP	
	"Maymkench"	F2	NP*	F0	INT
Males	"Safi"	F0	NP	INT	
	"Maymkench"	F0	F3	NP	INT

(NB: * the order depends on classifiers)

b. General classification models

As a second approach, we used all the features in the process of classification; we remarked that the recognition accuracy increases in a very noticeable way.

From Fig. 9, it can be seen that:

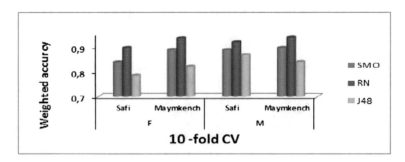

Fig. 9. Different weighted accuracies using all features for the three classifiers SMO, RN, and J48 according to speaker gender and phonemes.

- The best precisions are obtained for the word "Maymkench" (0.93 supplied by RN) which implies a dependence on phonemes.
- The recognition of the male's emotion is higher for the word "Safi" compared to that of females.

- The best rates are obtained by the algorithm RN whatever the gender of the speaker and whatever the phoneme.

In other hand, by examining the confusion matrices for women in Tables 2 and 3 and those for men in Tables 4 and 5 we noted that:

Table 2. Confusion matrix for the three algorithms for the word «Safi» (females).

SMO	A	N	F	Roc Area
A	48	0	6	0,932
N	2	47	5	0,873
F	6	20	28	0,711

RN	A	N	F	Roc Area
A	49	1	4	0,963
N	2	45	7	0,9
F	3	17	34	0,823

J48	A	N	F	Roc Area
A	37	4	13	0,805
N	6	36	12	0,804
F	9	8	37	0,746

Table 3. Confusion matrix for the three algorithms for the word «Maymkench» (females).

SMO	A	N	F	Roc Area
A	46	2	6	0,922
N	2	50	2	0,927
F	7	10	37	0,814

RN	A	N	F	Roc Area
A	48	2	4	0,961
N	0	48	6	0,951
F	6	8	40	0,891

J48	A	N	F	Roc Area
A	39	2	13	0,811
N	2	46	6	0,903
F	7	14	33	0,748

Table 4. Confusion matrix for the three algorithms for the word «Safi» (males).

SMO	A	N	F	Roc Area
A	47	0	7	0,913
N	0	51	3	0,962
F	11	7	36	0,787

RN	A	N	F	Roc Area
A	46	0	8	0,93
N	0	51	3	0,98
F	15	7	32	0,845

J48	A	N	F	Roc Area
A	51	0	3	0,919
N	0	47	7	0,907
F	7	9	38	0,778

Table 5. Confusion matrix for the three algorithms for the word «Maymkench» (males).

SMO	A	N	F	Roc Area
A	40	1	130	0,872
N	1	53	0	0,986
F	12	0	42	0,83

RN	A	N	F	Roc Area
A	43	1	10	0,906
N	1	53	0	0,984
F	12	0	42	0,921

J48	A	N	F	Roc Area
A	40	1	13	0,727
N	0	53	1	0,982
F	16	0	38	0,811

- Anger is best classified regardless phonemes and speaker gender.
- The area under the ROC curve is higher for anger.
- The RN and SMO algorithms have the best recognition rates which are almost identical.

5 Conclusion

In this paper, we studied the effect of phonemes and the speaker gender in the detection of emotions. The acoustic indices we have tested are: F1, F2, F3, F4, F0, Intensity, Number of Pulses, Jitter and Shimmer. We found that F0, INT, NP are the features that present a common behavior regardless phonemes and speaker gender. The second and third formant are the features that highlighted phoneme's effect. Indeed, the recognition rate increased for the word «Maymkench». In female's case, this increase is observed in F2 while in male's case, it is observed in F3. This may be explained by the explosion induced by the plosive consonant K and the characteristics of vocal tract related to speaker gender. Moreover, the best detected emotional state is anger for the two words independently of speaker gender. This result is provided by the three algorithms. This is may be due to the difficulty of expressing the feeling of fear by males which affected the quality of the data. The analysis of the area under the ROC curve confirmed the results. Comparison of the performances of three algorithms shows that RN and SMO are more competitive than J48.

References

1. Aggarwal, R.K., Dave, M.: Performance evaluation of sequentially combined heterogeneous feature streams for Hindi speech recognition system. Telecommun. Syst. **52**(3), 1457–1466 (2013)
2. Albornoz, E.M., Milone, D.H., Rubner, H.L.: Spoken emotion recognition using hierarchical classifiers. Comput. Speech Lang. **25**(3), 556–570 (2011)
3. Banse, R., Scherer, K.R.: Acoustic profiles in vocal emotion expression. J. Personal. Soc. Psychol. **70**(3), 614–636 (1996)
4. Batliner, A., Schuller, B., Seppi, D., Steidl, S., Devillers, L., Vidrascu, L., Vogt, T., Aharonson, V., Amir, N.: The automatic recognition of emotions in speech. In: Emotion-Oriented Systems, pp. 71–99. Springer, Heidelberg (2011)
5. Boersma, P.: Praat, a system for doing phonetics by computer. Glot Int. **5**(9/10), 341–345 (2001)
6. Burkhardt, F.: Simulation emotionaler Sprechweise mit Sprachsynthese verfahren. Ph.D. thesis, TU Berlin (2001)
7. Busso, C., Narayanan, S.S.: Interrelation between speech and facial gestures in emotional utterances: a single subject study. IEEE Trans. Audio Speech Lang. Process. **10**(20), 1–16 (2007)
8. Busso, C., Narayanan, S.S.: Joint analysis of the emotional fingerprint in the face and speech: a single subject study. In: International Workshop on Multimedia Signal Processing (MMSP), Chanée, Grèce, pp. 43–47. IEEE, Octobre 2007

9. Wu, C.H., Liang, W.B.: Emotion recognition of affective speech based on multiple classifiers using acoustic-prosodic information and semantic labels. IEEE Trans. Affect. Comput. **2**(1), 1–21 (2012)

10. Clore, G.L.: Why emotions are felt. In: Ekman, P., Davidson, R.J. (eds.) The Nature of Emotion: Fundamental Questions, pp. 103–111. Oxford University Press, New York (1994)

11. Cowie, R., Douglas-Cowie, E., Tsapatsoulis, N., Votsis, G., Kollias, S., Fellenz, W., Taylor, J.G.: Emotion recognition in human-computer interaction. IEEE Signal Process. Mag. **18**(1), 32–80 (2001)

12. Damasio, A.: L'erreur de Descartes. Grosset/Putnam, New York (1994)

13. Davletcharova, A., Sugathan, S., Abraham, B., James, A.P.: Detection and analysis of emotion from speech signals. Procedia Comput. Sci. **58**, 91–96 (2015)

14. Ekman, P.: Expression and the nature of emotion. In: Scherer, K.R., Ekman, P. (eds.) Approaches to Emotion, pp. 319–343. Lawrence Erlbaum Associates, Hillsdale (1984)

15. Hall, M., Frank, E., Holmes, G., Pfahringer, B., Reutemann, P., Witten, I.H.: The WEKA data mining software: an update. SIGKDD Explor. **11**(1), 10–18 (2009)

16. Hartmann, K., Siegert, I., Philippou-Hübner, D., Wendemuth, A.: Emotion detection in HCI: from speech features to emotion space. In: 12th IFAC Symposium on Analysis, Design, and Evaluation of Human-Machine Systems, Las Vegas, NV, USA, 11–15 August 2013

17. Hosmer, D.W., Lemeshow, S., Sturdivant, R.X.: Applied Logistic Regression, 3rd edn. Wiley, Hoboken (2013). ISBN 978-0470-58247-3

18. Huang, C., Gong, W., Fu, W., Feng, D.: A research of speech emotion recognition based on deep belief network and SVM. Math. Probl. Eng. (2014)

19. Johnstone, T., Scherer, K.R.: Vocal communication of emotion. In: Lewis, M., Haviland-Jones, J.M. (eds.) Handbook of Emotions, pp. 220–235. Guilford, New York (2000)

20. Juslin, P.N., Laukka, P.: Communication of emotions in vocal expression and music performance: different channels, same code. Psychol. Bull. **129**(5), 770–814 (2003)

21. Koolagudi, S.G., Rao, K.S.: Emotion recognition from speech: a review. Int. J. Speech Technol. **15**, 99–117 (2012)

22. Mower, E., Matarić, M.J., Narayanan, S.: A framework for automatic human emotion classification using emotion profiles. IEEE Trans. Audio Speech Lang. Process. **19**(5), 1057–1070 (2011)

23. Murray, I.R., Arnott, J.L.: Toward the simulation of emotion in synthetic speech: a review of the litterature on human vocal emotion. J. Acoust. Soc. Am. **93**(2), 1097–1108 (1993)

24. Oudeyer, P.: The production and recognition of emotions in speech: features and algorithms. Int. J. Hum.-Comput. Stud. **59**(1–2), 157–183 (2003)

25. Paeschke, A., Sendlmeier, W.: Prosodic characteristics of emotional speech: measurements of fundamental frequency movements. In: Speech Emotion, pp. 75–80 (2000)

26. Philippou-Hubner, D., Vlasenko, B., Bock, R., Wendemuth, A.: The performance of the speaking rate parameter in emotion recognition from speech. In: Proceedings of IEEE ICME, pp. 248–253 (2012)

27. Scherer, K.R.: Vocal affect expression: a review and a model for future research. Psychol. Bull. **99**(2), 143–165 (1986)

28. Scherer, K.R.: How emotion is expressed in speech and singing. In: Proceedings of 1995 ICPhS, Stockholm, pp. 90–96 (1995)

29. Stibbard, R.: Vocal Expression of Emotions in Non-Laboratory Speech: An Investigation of the Reading/Leeds Emotion in Speech Project Annotation Data, 245 p. Linguistics and Applied Language Studies, University of Reading, Reading, RoyaumeUni (2001)

30. Nwe, T.L., Foo, S.W., De Silva, L.C.: Detection of stress and emotion in speech using traditional and FFT based log energy features. In: Proceedings of the 4th International Conference on Information, Communications and Signal Processing (2009)
31. Vlasenko, B., Prylipko, D., Philippou-Hubner, D., Wendemuth, A.: Vowels formants analysis allows straightforward detection of high arousal acted and spontaneous emotions. In: Proceedings of INTERSPEECH 2011, Florence, Italy, pp. 1577–1580 (2011)
32. Vogt, T., Andre, E.: Comparing feature sets for acted and spontaneous speech in view of automatic emotion recognition. In: IEEE International Conference on Multimedia and Expo, pp. 474–477 (2005)
33. Yüncü, E., Hacıhabiboğluy, H., Bozşahin, C.: Automatic speech emotion recognition using auditory models with binary decision tree and SVM. In: Proceedings of the 2014 22nd International Conference on Pattern Recognition, pp. 773–778. Computer Society Washington, D.C. (2014). ISBN 978-1-4799-5209-0

Using Multiple Minimum Support
to Auto-adjust the Threshold of Support
in Apriori Algorithm

Azzeddine Dahbi[✉]⬤, Youssef Balouki, and Taoufiq Gadi

Laboratory Informatics, Imaging and Modelling of Complex Systems (LIIMSC),
Faculty of Science and Technology, Hassan 1st University Settat,
Settat, Morocco
azdine.im@gmail.com

Abstract. Nowadays, Data mining becomes an important research domain, aiming to extract the interesting knowledge and pattern from the large databases. One of the most well-studied data mining tasks is association rules mining. It discovers and finds interesting relationships or correlations among items in large databases. A great number of algorithms have been proposed to generate the association rules, one of the main problems related to the discovery of these associations (that a decision maker faces) is the choice of the threshold of the minimum support because it influences directly the number and the quality of the discovered patterns. To bypass this challenge, we propose in this paper an approach to determine how to auto-adjust the minimum support threshold according to data by using a multiple minimum support. The experiments performed on benchmark datasets show a significant performance of the proposed approach.

Keywords: Data mining · Association rules mining · APRIORI algorithm
Multiple minimum support

1 Introduction

Data mining technologies aim to explore a valuable knowledge in large volumes of data [1]. There are many data mining methods and algorithms, One of the most traditional data mining approaches is finding frequent item-sets in transactional databases, and deduct their corresponding association rules. Currently, there are many proposed algorithms for mining association rules. The most known and the simplest one is the APRIORI Algorithm [2] proposed in 1993 by Agrawal. The use of APRIORI algorithm in DM makes it possible to test the various combinations between the items (Data_Atributes) to find potential relationships which will be exposed in the form of association rules. However, the rules produced by APRIORI are judged by known measures (support and confidence). But this algorithm suffers from an important defect which cannot determine the minimal value of support and confidence and these parameters are estimated intuitively by the users. Depending on the choice of those thresholds, association rule mining algorithms can generate a huge number of rules

© Springer International Publishing AG, part of Springer Nature 2018
A. Abraham et al. (Eds.): SoCPaR 2017, AISC 737, pp. 111–119, 2018.
https://doi.org/10.1007/978-3-319-76357-6_11

which lead algorithms to suffer from long execution time and large memory consumption, or may generate a small number of rules, and thus may delete valuable information.

This method can also offers several rules in a massive database, millions, which apparently many of them are not useful and helpful; it can be implied that it doesn't have enough efficiency. So we require a method to find the best values of support parameter automatically especially in large databases. The main goal of this paper is to present a method to find proper values of minimum threshold for efficient support.

The outline of our paper is as follows: In Sect. 2, we present the necessary scientific background and an overview of association rules mining, and related works. Part 3 presents our proposed approach based on APRIORI algorithm for mining association rules with auto-adjust the threshold of support with multiple minimum support. In Sect. 4, we discuss the experimental results and its analysis. The conclusion and scope for future work is given in the last section.

2   An Overview of AR Mining

2.1   Process Association Rules Mining

Association Rules Mining

We define $I = \{i_1, i_2, \ldots\ldots i_n\}$ as a set of all items, and $T = \{t_1, t_2, \ldots\ldots t_m\}$ as a set of all transactions, every transaction t_i is an itemset and meets $t_i \subseteq I$. Association rules can be generated from large (frequent/closed/maximal) itemsets. An association rule is an implication expression of the form $X \rightarrow Y$, $X \subseteq I$, $Y \subseteq I$ where X and Y disjoint itemsets (i.e. $X \cap Y = \emptyset$). X is called the antecedent and Y is called the consequent of the rule.

The force of an association rule can be measured in terms of its support and confidence. The support of the rule $X \rightarrow Y$ is the percentage of transactions in database D that contain $X \cup Y$ and is represented as:

$$\text{Support}(X \rightarrow Y) = P(XY) = n(XUY)/n \tag{1}$$

The confidence of a rule $X \rightarrow Y$ describes the percentage of transactions containing X which also contain Y and is represented as

$$\text{Confidence}(X \rightarrow Y) = n(XUY)/n(X) = P(XY)/P(X) \tag{2}$$

Where $n(X \cup Y)$ is the number of transactions that contain items (i.e. XUY) of the rule, $n(X)$ is the number of transactions containing itemset X and n is the total number of transactions.

The process of mining association rules is to discover all association rules from the transactional database D that have support and confidence greater than threshold predefined by the user minimum support (minsup) and minimum confidence (minconf).

APRIORI Algorithm

Now, diverse algorithms for mining association rules are proposed. The most known, and without certainly the simplest one is the APRIORI algorithm [2]. It scans the mesh of the concepts width, such as Charm [3] and Closet [4] algorithms. Other travel the lattice depth is particularly the case for algorithms FP-Growth [5] and Eclat [6].

The APRIORI algorithm works in two steps:

- Find the frequent itemset: The frequent itemset is an itemset that verifies a predefined threshold of minimum support.
- Generate all strong association rules from frequent itemsets: The strong association rule is a rule that verifies a predefined threshold of minimum confidence.

APRIORI algorithm is the most powerful method that candidate $k + 1$-itemsets may be generated from frequent k-itemsets according to the principle of APRIORI algorithm that any subset of frequent itemsets are all frequent itemsets.

Foremost, find the frequent 1-itemsets L1. Then L2 is generated from L1 and so on, until no more frequent k-itemsets can be found and then algorithm desists. Every Lk generated should scan database once. Then Ck is generated from Lk − 1.

Pseudo code of APRIORI:

```
Ck: candidate itemset of size k.
Lk: frequent itemset of size k.
L1: frequent items.
For (k = 1;Lk! = o;k ++) do begin
Ck+1 = candidate generated from Lk;
For each transaction t in database D do increment the
calculation of all candidates in Ck+1 that are included in
t.
Lk+1 = candidate in Ck+1 with minsup.
End.
Return U(Lk).
```

2.2 State of the Arts

The process of association rules mining (ARM) can be categorized into two classes of research: determination of user specified support and confidence thresholds and the post-treatment by using the interestingness measures to evaluate and find the most interesting rules. Most of the algorithms of ARM rely on support and confidence thresholds and they use a uniform threshold at all levels. Therefore a suitable choice of those thresholds directly influences the number and the quality of association rules discovered.

Several works aim to solve this challenge and help the user in the choice of the threshold of support and confidence to be the most adequate to the decision scope.

Fournier-Viger [7, 8] Redefine the problem of association rule mining as mining the top-k association rules by introducing an algorithm to find the top k rules having the greatest support. With k is the number of rules to be generated and defined by the users. To generate rules, Top-K-Rules relies on a novel approach called rule expansions, it finds larger rules by recursively scanning the database for adding a single item at a time to the left or right part of each rule. This has an excellent scalability: execution time linearly increases with k. Top-k pattern mining algorithm is slower but provides the benefit of permitting the user set the number of patterns to be discovered, which may be more intuitive.

Kuo et al. [9] introduced a new method to determine best values of support and confidence parameters automatically particularly for finding association rule mining using Binary Particle Swarm Optimization and offered a novel approach for mining association rule in order to develop computational performance as well as to automatically define suitable threshold values. The particle swarm optimization algorithm searches firstly for the best fitness value of each particle and then detects corresponding support and confidence as minimal threshold values after the data are converted into binary values and then these minimal support and confidence values are used to extract association rules.

Other approaches [10–12] used a multiple level in the process of ARM, the multiple level association rule mining can run with two kinds of support: Uniform and Reduced. Uniform Support: In this method, same minimum support threshold is used at every level of the hierarchy. There is no necessity to evaluate itemsets including items whose ancestors do not have minimum support. The minimum support threshold has to be suitable. If minimum support threshold is too great then we can lose lower level associations and if it is too low then we can end up in producing too many uninteresting high-level association rules.

Other works [13, 14] proposed an approach based on multi-criteria optimization aiming to select the most interesting association rules Without need to set any parameters at all, The idea is to find the patterns that are not dominated by any other patterns by using a set of interesting measures.

3   Proposed Approach

The main problem of The APRIORI algorithm is the choice of the threshold for the support and confidence. APRIORI find the frequent candidate itemset by generating all possible candidate itemset which verifies a minimum threshold defined by users. This choice influences in the number and the qualities of AR. whereas our algorithm uses a threshold of minsup defined depending on the transactional dataset which is logical.

In this paper we propose two main contributions, the first one is to compute the minimum support (minsup) automatiquelly according to each datasets instead of using a constant value predefined by the users. The second contribution of our proposed method is making this minsup change (updated) dynamically according to each level, most of the existing methods applied a single and uniform minimum support threshold value for all the items or itemsets. But all the items in an itemset do not work in the same process, some appear very frequently and oftentimes, and some infrequent and very rare. Therefore the threshold of minsup should change according to different levels of itemset.

Our algorithm can be divided in several steps:

- Input: a set of n transaction, a transactional dataset.
- Step 1: determine the minimum support for the first level for 1-itemset: minsup1 by using the means of support of all itemset with one item.

$$\min \sup 1 = \sum_{i=1}^{N} \frac{\sup - 1 itemset_i}{N}$$

Minsup: is a minimum support and 1-itemset is a set of items composed of 1item.

- Step 2: Verify whether the support sup − 1itemset$_i$ of each item$_i$ is large than or equal to minsup1. If i satisfies the above condition put in the set of 1-itemset (L$_1$). L1 = {1-itemset$_i$/sup − 1-itemset$_i$ ≥ minsup1 with i = 1…..N number of all 1-itemset}
- Step 3: Generate the candidate C$_2$ from L$_1$ with the same way to the APRIORI algorithm. the difference is that the support of all the large k-itemset. k-itemset: set of items composed of k items.
- Step 4: Compute the new minsup of the 2 itemset level by using the means of support of the generated C$_2$ generate the L$_2$. L$_2$ = {2-itemset$_i$/sup − 2-itemset$_i$ ≥ minsup2 with i = 1……N number of all 2-itemset}
- Step 5: Check whether the support sup k-itemset$_i$ of each candidat k-itemset$_i$ is larger than or equal to minsup k obtained in step 4. If it satisfies the above condition put in the set of large k-itemset(L$_k$). L$_k$ = {k-itemset$_i$/sup − kitemset$_i$ ≥ minsupk with i = 1….N number of all kitemset}.
- Step 6: Repeat steps 3 to 5 until L$_i$ is null.
- Step 7: Construct the association rules for each large k-itemset$_i$ with items: {Ik1, Ik2,……. Ikq} q ≥ 2 which verify the threshold of confidence i.e. the association rule whose confidence values are larger than or equal to the threshold of confidence defined by the mean of support of all large q itemset Ik.
- Output: a set of association rules using an automatic threshold of support in multilevel.

4 Experiment Study

In this part, we will illustrate and investigate the advantages of our proposed algorithm (Supd), we use different public datasets: (mushroom, flare1, flare2, Zoo, Connect) got from UCI machine learning repository [15]. T10I4D100K (T10I4D) was generated using the generator from the IBM Almaden Quest research group [16] and Foodmart is a dataset of customer transactions from a retail store, obtained and transformed from SQL-Server 2000 [17]. Table 1 summarizes the properties of the used datasets.

Our objectives in this section are multiple, the first, we show through many experiments that our method reduce the huge number of the generated association rules compared to APRIORI algorithm [2]. Second, we conduct an experiment to examine the qualities of the generated association rules. The third, we study the runtime and we compare it to the APRIORI algorithm (APR) and to Topkrule algorithm (Topk) [7].

All the approaches are implemented in Java programing language. At the first, all our experiments are realized through a computer (C1) with the following specifications: Core™ I3, 1,70 GHz, memory capacity: 4 GB.

Table 1. Characteristics of the used datasets

Data set	Items	Transactions
Mushroom	22	8124
Flare1	32	323
Flare2	32	1066
Zoo	28	101
Connect	42	67557
Chess	36	3196
Foodmart	1,559	4141
T10I4D	1000	100000

Table 2 shows different values of threshold of confidence chosen and different values of support found by our algorithm in the first level for different dataset.

Table 2. Threshold values of support and confidence

Data set	Minsup	Minconf
Mushroom	20	70
Flare1	32	50
Flare2	32	50
Zoo	30	50
Connect	33	80
Chess	76	70
Foodmart	0.02	70
T10I4D	0.2	70

4.1 Reduction of a Number of Rules

In this experiment, we show the ability of our proposed approach to reduce the number of AR generated from the chosen datasets. Our experiment compares our approach to APRIORI based on thresholds.

Table 3 compares the size of AR generated by our method to the APRIORI.

Table 3. Number of AR generated for each dataset

	Mushroom	Flare1	Flare2	Zoo	Connect	Chess	Foodmart	T10I4D
Supd	8353	106	127	362	422874	55700	2037	111482
APR	14965927	1430	1298	3716	220224860	1776698	21982	206772

We see through this experiments that our approach can significantly reduce the huge number of rules generated from the data sets, which can facilitate the interpretation and help the users to see the most interesting ones and to take decision.

4.2   The Running Time Analysis

We realized an implementation for traditional Apriori from [1] and our proposed algorithm (Supd), and we compare the time wasted of original Apriori (APR) and Topkrule algorithm (Topk), and our proposed algorithm by applying many datasets, various values for the minimum support given in the implementation. The running time analysis may be differ for different machine configuration. For this reason we are realized our experiments through another machine computer (C2) with the following specifications: Core™2 Duo CPU E8400, 3,00 GHz, memory capacity: 4 GB in order to obtain unbiased result comparison. The result is shown in Tables 4 and 5.

As we see in Table 4, that the time-consuming in our proposed algorithm in each dataset is less than it is in the original Apriori, and the difference grows more and more as the number of transactions of datasets increases.

On the other hand we see that this time consuming in our approach is the same as the time consuming in Topkrule in some datasets and it is less than it in other datasets.

We can add another advantage to our algorithm which is the use of memory space. As we see, we did not obtain the result of Topkrule in Connect and T10I4D100K datasets, the Topkrule algorithm can't run on both machines and this is due to the memory problem, while our algorithm runs without any problem.

Table 4. The time consuming in different datasets using computer C1 (in ms)

	Muchroom	Flare1	Flare2	Zoo	Connect	Chess	Foodmart	T10I4D
Supd	3746	10	73	24	224335	1852	20660	1283038
APR	86984	100	81	60	888383	9432	62152	340642
Topk	11419	10	20	12	–	41501	18600	–

Table 5. The time consuming in different datasets using computer C2 (in ms)

	Muchroom	Flare1	Flare2	Zoo	Connect	Chess	Foodmart	T10I4D
Supd	5728	10	83	24	336146	2037	24309	166599
APR	96935	100	81	60	1557284	10500	84201	689908
Topk	19759	20	43	15	–	52978	4968	–

4.3 The Quality of the Extracted Rules

In order to analyze the performance of our proposed algorithm, we have compared the average value of confidence in each dataset of our method to the original method.

The Table 6 shows that the proposed method has found rules with high values of confidence in the majority of the datasets which ensures the benefit of our proposed method.

Table 6. The average of confidence for different datasets

	Muchroom	Flare1	Flare2	Zoo	Connect	Chess	Foodmart	T10I4D
Supd	0.89	0.85	0.9	0.86	0.96	0.94	1	0.87
APR	0.80	0.75	0.85	0.84	0.84	0.86	0.98	0.78
Topk	0.89	0.85	0.9	0.88	–	0.94	0.99	–

5 Conclusion

In this paper, we proposed a new approach based on APRIORI algorithm for discovering the association rules to auto-adjust the choice of the threshold of support. The main advantage of the proposed method is the automatism of the choice of support in multi-level, we get results containing desired rules with maximum interestingness in a little time. The numbers of rules generated by proposed algorithm are significantly less as compared to APRIORI Algorithm. Hence, we can say our algorithm answer the problem of the choice of the threshold of the support efficiently and effectively. As future works, we plan to ameliorate our approach to be able to select the interesting association rules without using any predefined threshold.

References

1. Zhou, S., Zhang, S., Karypis, G. (eds.): Advanced Data Mining and Applications: 8th International Conference, ADMA 2012, 15–18 December 2012, Proceedings, vol. 7713. Springer Science & Business Media, Nanjing (2012)
2. Agrawal, R., Srikant, R.: Fast algorithms for mining association rules. In: Proceedings of VLDB 1994 Proceedings of 20th International Conference on Very Large Data Bases, vol. 1215, pp. 487–499 (1994)
3. Zaki, M.J., Hsiao, C.J.: CHARM: an efficient algorithm for closed itemset mining. In: SDM 2002, Arlington, VA, pp. 457–473 (2002)
4. Pei, J., Han, J., Mao, R.: CLOSET: an efficient algorithm for mining frequent closed itemsets. In: Proceeding of the 2000 ACM-SIGMOD International Workshop Data Mining and Knowledge Discovery (DMKD 2000), Dallas, TX, pp. 11–20. ACM (2000)
5. Pei, J., Han, J., Mortazavi-Asl, B., Wang, J., Pinto, H., Chen, Q., Dayal, U., Hsu, M.-C.: Mining sequential patterns by pattern-growth: the prefixspan approach. IEEE Trans. Knowl. Data Eng. **16**, 1424–1440 (2004)

6. Schmidt-Thieme, L.: Algorithmic features of Eclat. In: Proceedings of the IEEE ICDM Workshop on Frequent Itemset Mining Implementations (FIMI 2004), CEUR Workshop Proceedings, Brighton, UK, vol. 126 (2004)
7. Fournier-Viger, P., Wu, C.-W., Tseng, V. S.: Mining top-k association rules. In: Proceedings of the 25th Canadian Conference on Artificial Intelligence (AI 2012), pp. 61–73. Springer, Canada (2012)
8. Fournier-Viger, P., Tseng, V.S.: Mining top-k non-redundant association rules. In: Proceedings of 20th International Symposium, ISMIS 2012, LNCS, Macau, China, vol. 7661, pp. 31–40. Springer (2012)
9. Kuo, R.J., Chao, C.M., Chiu, Y.T.: Application of particle swarm optimization to association rule mining. In: Proceeding of Applied Soft Computing, pp. 326–336. Elsevier (2011)
10. Liu, B., Hsu, W., Ma, Y.: Mining association rules with multiple minimum supports. In: Knowledge Discovery and Databases, pp. 337–341 (1999)
11. Lee, Y.C., Hong, T.P., Lin, W.Y.: Mining association rules with multiple minimum supports using maximum constraints. Int. J. Approx. Reason. **40**(1), 44–54 (2005)
12. Hu, Y.-H., Chen, Y.-L.: Mining association rules with multiple minimum supports: a new algorithm and a support tuning mechanism. Decis. Support Syst. **42**(1), 1–24 (2006)
13. Bouker, S., Saidi, R., Ben Yahia, S., Mephu Nguifo, E.: Mining undominated association rules through interestingness measures. Int. J. Artif. Intell. Tools **23**(04), 1460011 (2014)
14. Dahbi, A., Jabri, S., Ballouki, Y., Gadi, T.: A new method to select the interesting association rules with multiple criteria. Int. J. Intell. Eng. Syst. **10**(5), 191–200 (2017)
15. UCI machine learning repository. https://archive.ics.uci.edu/ml/index.php. Accessed 10 Jan 2018
16. Frequent Itemset Mining Implementations Repository. http://fimi.ua.ac.be/data/. Accessed 10 Jan 2018
17. An Open-Source Data Mining Library. http://www.philippe-fournier-viger.com/spmf/index.php?link=datasets.php. Accessed 10 Jan 2018

Prediction of the Degree of Parkinson's Condition Using Recordings of Patients' Voices

Clara Jiménez-Recio[1], Alexander Zlotnik[1], Ascensión Gallardo-Antolín[2(✉)], Juan M. Montero[1], and Juan Carlos Martínez-Castrillo[3]

[1] Speech Technology Group, ETSIT, Universidad Politénica de Madrid, Avda. Complutense, 30, 28040 Madrid, Spain
juancho@die.upm.es
[2] Department of Signal Theory and Communications, Universidad Carlos III de Madrid, Avda. de la Universidad, 30, 28911 Leganés, Madrid, Spain
gallardo@tsc.uc3m.es
[3] Service of Neurology, Hospital Universitario Ramón y Cajal, Ctra. de Colmenar Viejo, km. 9.100, 28034 Madrid, Spain
jcmcastrillo@gmail.com

Abstract. This paper addresses the estimation of the degree of Parkinson's Condition (PC) using exclusively the patient's voice. Firstly, a new database with speech recordings of 25 Spanish patients with different degrees of PC is presented. Secondly, we propose to face this problem as a regression task using machine learning techniques. In particular, utilizing this database, we have developed several systems for predicting the PC degree from a set of acoustic characteristics extracted from the patients' voice, being the most successful ones, those based on the Support Vector Regression (SVR) algorithm. To determine the optimal way of exploiting the data for our purposes, three kind of experiments have been considered: cross-speaker, leave-one-out-speaker and multi-speaker. From the results, it can be concluded that prediction systems based on acoustic features and machine learning algorithms can be applied for tracking the PC progression if enough validation/training speech data of the patient is available.

Keywords: Parkinson's disease · UPDRS · Speech features
Machine learning · Regression · SVR

1 Introduction

Parkinson's Disease (PD) is a chronic progressive neurodegenerative disorder of the central nervous system that causes tremors and partial or full loss in motor reflexes, coordination, muscle stiffness, speech, behavior, mental processing, and other vital functions [1]. It is estimated that Parkinson's Condition (PC) affects

© Springer International Publishing AG, part of Springer Nature 2018
A. Abraham et al. (Eds.): SoCPaR 2017, AISC 737, pp. 120–129, 2018.
https://doi.org/10.1007/978-3-319-76357-6_12

from 7 to 10 million people worldwide (0.1%), about 80% of them show some form of vocal impairment [2]. The monitoring of this disease progression, which is usually done through the so-called Unified Parkinson's Disease Rating Scale (UPDRS) and is performed by expert medical staff, is hard and costly. Besides, it requires the physical presence of the patient, what is sometimes troublesome. The idea of using speech tests arises as a solution to these problems as is a fast and non-invasive method that allows remote monitoring and feedback.

This paper focuses on the development of a system to automatically determine the degree of PC using recordings of patients' voices and machine learning techniques. Since this disease affects not only speech but also other important aspects of human health, tackling the analysis of PC progression using only the patients' voice is a complicated issue [3]. In addition, there seems to be a strong relationship between PC progression and speech degradation [4].

First studies dealing with the problem of automatically tracking the PC progression through speech, used only sustained vowel recordings from which a set of dysphonia measurements were extracted [5]. However, as pointed in [6], the analysis of prosody and intelligibility requires the inclusion of running speech. In this context, in the work by [7], in which pitch-related cues were also considered, it is shown that a reading task is better for automatically assessing the severity of PC than sustained phonation tasks. The system developed in [8] also deals with different types of data (sustained vowels, words, running speech,...) and includes acoustic, automatic speech recognition and intelligibility features and a regressor based on the Random Forest algorithm.

Our work goes into two main directions. Firstly, a new database with speech recordings of 25 Spanish patients with different degrees of PC is presented. Secondly, we propose a system for UPDRS prediction based on acoustic characteristics extracted from patients' voice and regression machine learning algorithms. We have taken into consideration two different scenarios. In the first one, the system is applied on a totally new unknown patient in order to estimate his/her UPDRS. In the second one, the system is intended to be used for tracking the PC evolution of a known patient.

The remainder of this paper is organized as follows: Sect. 2 describes the database recorded and used in our experimentation. Section 3 describes the prediction system, the acoustic features and regression techniques considered for this task. Our experiments and results are presented in Sect. 4, followed by some conclusions of the research in Sect. 5.

2 Database

We have collected a specific audio database for our experiments whose design is based on the structure of the PC speech database described in [6]. It consists of speech data from 25 Spanish native participants diagnosed with Parkinson's disease with several degrees of severity and was recorded with a head-mounted microphone at 48 kHz with the patients in the ON-state and in a quiet room. Subjects were asked to explicitly grant permission for the use of their data for this research. Each participant produced a total of 49 audio files in Spanish with the following distribution: three repetitions of the five Spanish vowels uttered in a sustained manner, 25 common words, 8 sentences and a free monologue. As the monologues have not been used in our experimentation due to their distinctive nature, only 48 speech files per speaker are available.

After the recording session, medical experts measured the UPDRS of each patient. Additional information as the genre, age, time from the first diagnosis and the Hoehn and Yahr (H&Y) scale is also available. The H&Y describes how the symptoms of PC progresses and it includes stages 1 (unilateral involvement only) through 5 (wheelchair bound or bedridden unless aided), with the addition of stages 1.5 and 2.5. Table 1 contains the demographics information of the participants in the database.

Table 1. Demographics information of the speakers in the database

Subject	Gender	Age	H&Y	UPDRS	Subject	Gender	Age	H&Y	UPDRS
1	M	70	2	25	14	M	68	2	50
2	M	64	2	69	15	F	65	2	55
3	M	64	2.5	49	16	M	80	5	87
4	F	67	2.5	66	17	M	66	2	24
5	M	78	2	42	18	F	70	1	23
6	M	67	2	27	19	M	75	3	54
7	M	63	2	18	20	F	79	4	59
8	F	50	2	38	21	M	69	2	21
9	F	77	2.5	52	22	M	54	2	22
10	F	57	2	56	23	M	73	2	19
11	M	68	3	57	24	M	57	2	6
12	M	59	4	74	25	F	45	2	48
13	M	62	2	71					

3   Automatic PC Progression Prediction System

The automatic prediction of the degree of Parkinson's condition using only the information contained in patients' voice can be formulated as a machine learning problem consisting of the following two primary stages: feature extraction and regressor or predictor. The first stage obtains a parametric and compact representation of the speech signals that are more appropriate for the task. The purpose of the second one is to determine the PC degree (as measured according to the UPDRS) that corresponds to the analyzed speech signal using a certain regression process. In next subsections, we detail the main characteristics of both modules.

3.1 Feature Extraction

The set of acoustic features considered in this work corresponds to the one proposed in the PC Subchallenge of the INTERSPEECH 2015 Computational Paralinguistics Challenge [3]. They have been extracted using the feature extraction toolbox *openSMILE* version 2.1 [9] and consist of a set of 6373 utterance-level acoustic features computed over low-level descriptors such as energy, Mel-Frequency Cepstrum Coefficients (MFCC), pitch and voice quality features as jitter, shimmer and Harmonic-to-Noise Ratio (HNR), etc. These features cover phonetic, articulatory and prosodic characteristics of the impaired speech.

3.2 Predictor

In this Section, we briefly describe the machine learning algorithms utilized in the regression module.

Linear Regression. Linear Regression (LR) predicts a real-valued output based on an input value that can be defined by multiple input features or attributes. Linear models stand out for their simplicity; they can solve classification problems where data are composed of numerical features. The disadvantage of these algorithms is, indeed, the impossibility of creating nonlinear decision boundaries between classes. This aspect makes them too simple for many applications.

Support Vector Regression. Support Vector Machine (SVM) is a well-known supervised classification technique that uses linear models to implement nonlinear class boundaries [10] by transforming the input features to a new space by means of a nonlinear mapping. Support Vector Regression (SVR) [11] is basically an extension of SVM to solve regression tasks. SVM/SVR is considered by many to be one of the most powerful "black box" learning algorithms. For the implementation of the SVR-based regressors we have used the open-source machine learning Python toolkit *scikit-learn* version 0.16.0 [12].

Random Forest. A Random Forest (RF) is an ensemble of regression trees trained from randomly-sampled sets of data, sharing a common distribution [13]. The algorithm first creates a "bag" of samples by random sampling from the training set, then creates a tree-based ranker for each "bag" of data, and finally ensembles the full forest of trees. In this paper, we have used an open-source implementation of Random Forest, *RankLib* version 2.1, included in the Lemur project [14]. In particular, we have utilized the variant of RF with Multiple Additive Regression Trees (MART).

Lasso and Elastic Net. The Lasso (Least Absolute Shrinkage and Selection Operator) is a regression linear model with regularization that estimates sparse coefficients, in such a way that reduces the number of attributes upon which the given solution is dependent [15]. In some of our experiments, we have employed Elastic Net, a Lasso extension, for performing feature selection. In particular, we have used the implementation of Elastic Net included in the open-source machine learning Python toolkit *scikit-learn* version 0.16.0 [12].

4   Experiments

We have performed three different types of experiments (cross-speaker, leave-one-out-speaker and multi-speaker) with the aim of determining the best way to use the available speech data in order to obtain the most accurate results, i.e. the best prediction of the UPDRS state of a given patient. The first kind of experiment is related to the scenario where the system is applied on a totally new unknown patient, whereas the two remainder ones refer to the situation where the system is intended to be used for monitor the PC progress of a known patient. In next subsections, we detail these experiments and the corresponding results.

In all cases, the system outputs a UPDRS prediction per audio file. These predictions are subsequently averaged per speaker as the ultimate goal of our application is to obtain an UPDRS assessment for each evaluated patient.

Following [3], the performance of the proposed systems is measured in terms of the Spearman's correlation coefficient computed between the reference UPDRS labels determined by a neurologist expert and the predicted UPDRS values per speaker produced by the different learning algorithms. In addition, we also report the relative error (RError) of the predictor system computed as,

$$RError = \frac{1}{N_{spk}} \sum_{i=1}^{N_{spk}} \frac{|UPDRS_{ref}(i) - UPDRS_{pred}(i)|}{UPDRS_{ref}(i)} * 100 \qquad (1)$$

where N_{spk} is the number of speakers in the database and $UPDRS_{ref}(i)$ and $UPDRS_{pred}(i)$ are respectively, the reference and predicted UPDRS labels for the i-th speaker.

4.1   Cross-Speaker Experiment

In the cross-speaker experiment, we intent to analyse the case where the system is going to be applied on unknown patients. For this purpose, we adopt a 5-fold cross-validation strategy. In particular, we made a division of the data into five disjoint groups of five patients each, trying that each group was as homogeneous as possible according to the genre, age and UPDRS distribution. In each fold, four groups are used for training (20 speakers), and the remainder one (5 speakers) is divided into two subgroups, employing half of the audio files per speaker for validation whereas the other half is kept for testing. This process is repeated 5 times until all groups have been used once for testing.

Table 2 shows the Spearman's coefficient and relative error achieved by the SVR and RF learning algorithms and their combination with Lasso. As mentioned before, in this work, Lasso is used as a feature selection technique prior to the application of SVR/RF, reducing, in this case, the original repertory of 6373 features to a new set of 403 attributes. On the one hand, as can be observed, SVR improves significantly the performance of RF with and without Lasso. For this reason, we have discarded RF-based systems for the rest of the experimentation.

On the other hand, although Lasso provides a significant reduction of computational load, it degrades the performance of the SVR-based system. Nevertheless, it produces a slight improvement in the RF-based one.

Similar experiments were repeated using only the audio files corresponding to sustained phonations of vowels. Unfortunately, results were not satisfactory, so it seems that it is preferable to work with as much variety and amount of speech data as possible.

Any case, results are poor and from them it is possible to conclude that it is necessary to have a priori some speech data of the patient to be evaluated for the development of the prediction system. For exploring this possibility, we have performed two new kind of experiments: leave-one-out speaker and multi-speaker as will be described in next subsections.

Table 2. Cross-speaker results

Method	Spearman	Relative error
SVR	0.344	69.06 %
RF	0.092	71.86 %
SVR+Lasso	0.296	69.38 %
RF+Lasso	0.121	69.58 %

4.2   Leave-One-Speaker-Out (LOSO) Experiment

With the leave-one-speaker-out methodology, we approach the case where some data of the speaker to be evaluated is available. This is a realistic situation if the system is intended to be applied for studying the PC symptoms evolution in a known patient due to several factors (medication changes, etc.).

In this experiment, the predictor is trained with 24 patients, optimised with half of the data of the remaining patient and tested with the other half data of this patient. The process is repeated 25 times until all speakers have been left out once and tested with the rest of the speakers. Figure 1 represents this methodology.

Table 3 shows the results with this strategy and SVR and SVR+Lasso. It can be observed that both the Spearman's coefficient and the relative error improve considerably with respect to the cross-speaker case. The best result is achieved by using Lasso to reduce the number of input attributes. Although the performance differences with respect to only SVR are not significant, SVR+Lasso decreases drastically the computational load of the system.

Ensemble Learning. Trying to refine the predictions produced by the LOSO strategy, we developed an ensemble of two regressors, in such a way that the predictions generated by the LOSO system could be used as input parameters (among others) of a second-level regressor. In particular, the input attributes

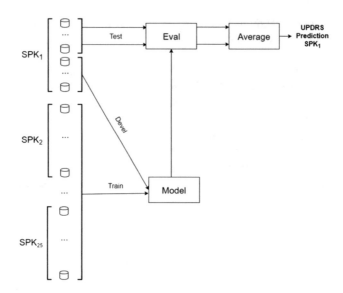

Fig. 1. Leave-one-speaker-out strategy

Table 3. Leave-one-speaker-out results

Method	Spearman	Relative error
SVR	0.513	58.3 %
SVR+Lasso	0.516	54.1 %

of this new regressor, which is trained using the conventional Linear Regression technique, are:

- UPDRS predicted for each patient by the LOSO system
- Hoenh & Yahr stage
- Age of the patient
- Time since the patient was diagnosed
- Genre of the patient

Unfortunately, the Spearman's coefficient achieved after this refinement was lower than the one produced by the LOSO system by itself. This deterioration could be caused by the low number of instances which the second-level predictor was trained with (24 instances of 5 attributes each $= 120$ instances) compared to 1,152 instances (24 patients with 48 audios each one) which the first-level (LOSO) regressor was trained with. Although this experiment has not been successful, it should be noted that the use of data, especially personal data, in applications of data mining and machine learning, acquires serious ethical implications [10]. Therefore, it is necessary to act responsibly with the data we handle. However, the situation is complex: everything depends on the application. Using this kind of information for medical diagnosis, as in the case at hand, is certainly ethical.

4.3   Multi-speaker Experiment

The idea behind this experiment is that speech data of the speaker to be evaluated is used not only for the predictor optimization (as in the previous case), but also for its training. In particular, a single predictor is trained for all speakers using 50% of the available data (24 audio files per each speaker), optimised with 25 % of the data (12 audio files per each speaker) and tested with the remaining 25 % (12 audio files per each speaker). This methodology is represented in Fig. 2. Finally, we got a system (SVR, $C = 0.01$, $L = 0.2$) trained, optimised and evaluated with audio of every patient.

Table 4 shows the results obtained with this strategy. The Spearman's coefficient and the relative error produced by the cross-speaker and LOSO methodologies are also included for comparison purposes. As can be observed, best results are achieved by the multi-speaker experiment with significant improvements in both, the Spearman's coefficient and the relative error, respect to cross-speaker and LOSO experiments.

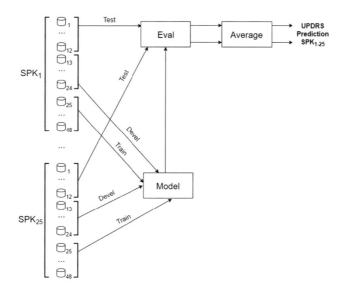

Fig. 2. Multi-speaker strategy

Table 4. Comparative results with the SVR+Lasso-based prediction system

Experiment	Spearman	Relative error
Cross-speaker	0.296	69.38 %
Leave-one-speaker-out	0.516	54.10 %
Multi-speaker	0.872	34.20 %

5   Conclusion

In this paper, we have addressed the problem of predicting the degree of Parkinson's condition, as measured by means of the UPDRS scale, in a patient by using exclusively his/her voice. Firstly, we have presented a new database in Spanish composed of 25 patients with different degrees of PC. Secondly, we have developed several systems based on regression machine learning algorithms for UPDRS prediction.

In order to determine the optimal way of exploiting the recorded speech data for our purposes, three kind of experiments have been carried out: cross-speaker, where no data of the patient to be evaluated (target speaker) is used for training/validation; leave-one-out-speaker, where part of the target speaker data is used for validation but not for training; and multi-speaker, where part of the target speaker data is used for training and validation. Results achieved by the cross-speaker strategy (i.e. when the system is used for predicting the UPDRS of an unknown speaker) are low, indicating that it is necessary to use some data of the target speaker for optimising and/or training the system. The leave-one-speaker-out and multi-speaker experiments try to reflect the scenario where the system is applied on a known patient in order to monitor the progression of the PC. A SVR+Lasso-based prediction system obtains a Spearman's coefficient of 0.516 in the LOSO strategy and of 0.872 in the multi-speaker one. From these results, it is possible to conclude that automatic systems based on acoustic features and machine learning algorithms can be applied for analysing the progression of PC in a patient if enough validation/training speech data of this speaker is available.

Acknowledgments. The work leading to these results has been partly supported by Spanish Government grants TEC2014-53390-P and DPI2014-53525-C3-2-R. Authors also thank all the patients and the medical staff of the Service of Neurology of the Hospital Universitario Ramón y Cajal involved in the speech data acquisition for their generous contribution to our research ("Estudio sobre clasificación automática de pacientes diagnosticados con Parkinson según la escala UPDRS usando la voz").

References

1. Jankovic, J.: Parkinson's disease: clinical features and diagnosis. J. Neurol. Neurosurg. Psychiatry **79**(4), 368–376 (2007)
2. Rusz, J., Cmejla, R., Ruzickova, H., Ruzicka, E.: Quantitative acoustic measurements for characterization of speech and voice disorders in early untreated Parkinsons' disease. J. Acoust. Soc. Am. **129**(1), 350–367 (2011)
3. Schuller, B., Steidl, S., Batliner, A., Hantke, S., Honig, F., Orozco-Arroyave, J.R., Noth, E., Zhang, Y., Weninger, F.: The INTERSPEECH 2015 computational paralinguistics challenge: nativeness, Parkinsons & eating condition. In: Proceedings of 16th Annual Conference of the International Speech Communication Association (INTERSPEECH 2015), Dresden, Germany (2015)
4. Skodda, S., Rinsche, H., Schlegel, U.: Progression of dysprosody in Parkinson's disease over time: a longitudinal study. Mov. Disord. **24**(5), 716–722 (2009)

5. Tsanas, A., Little, M.A., McSharry, P.E., Ramig, L.O.: Accurate telemonitoring of Parkinson's disease progression by noninvasive speech tests. IEEE Trans. Biomed. Eng. **57**(4), 884–893 (2010)
6. Orozco-Arroyave, J.R., Arias-Londoño, J.D., Vargas-Bonilla, J.F., Gonzalez-Rátiva, M.C., Nöth, E.: New Spanish speech corpus database for the analysis of people suffering from Parkinson's disease. In: Proceedings of Ninth International Conference on Language Resources and Evaluation (LREC 2014), pp. 342–347 (2014)
7. Bayestehtashk, A., Asgari, M., Shafran, I., McNames, J.: Fully automated assessment of the severity of Parkinson's disease from speech. Comput. Speech Lang. **29**(1), 172–185 (2015)
8. Zlotnik, A., Montero, J.M., San-Segundo, R. Gallardo-Antolín, A.: Random forest-based prediction of Parkinson's disease progression using acoustic, ASR and intelligibility features. In: Proceedings of 16th Annual Conference of the International Speech Communication Association (INTERSPEECH 2015), Dresden, Germany, pp. 503–507 (2015)
9. Eyben, F., Weninger, F., Gross, F., Schuller, B.: Recent developments in openSMILE, the Munich open-source multimedia feature extractor. In: Proceedings of ACM Multimedia (MM), pp. 835–838 (2013)
10. Hall, M., Witten, I., Eibe, F.: The Grid: Data Mining: Practical Machine Learning Tools and Techniques. Morgan Kaufmann Publishers, San Francisco (1999)
11. Drucker, H., Burges, C.J.C., Kaufman, L., Smola, A., Vapnik, V.: Support vector regression machines. In: Advances in Neural Information Processing Systems (NIPS 1996), pp. 155–161. MIT Press (1997)
12. Pedregosa, F., Varoquaux, G., Gramfort, A., Michel, V., Thirion, B., Grisel, O., Blondel, M., Prettenhofer, P., Weiss, R., Dubourg, V., Vanderplas, J., Passos, A., Cournapeau, D., Brucher, M., Perrot, M., Duchesnay, E.: Scikit-learn: machine learning in Python. J. Mach. Learn. Res. **12**, 2825–2830 (2011)
13. Breiman, L.: Random forests. Mach. Learn. **45**(1), 5–32 (2001)
14. Dang, V.: RankLib - a library of ranking algorithms. https://sourceforge.net/p/lemur/wiki/RankLib
15. Tibshirani, R.: Regression shrinkage and selection via the lasso. J. R. Stat. Soc. Ser. B (Methodol.) **58**(1), 267–288 (1996)

An Overview of a Distributional Word Representation for an Arabic Named Entity Recognition System

Chaimae Azroumahli$^{(\boxtimes)}$, Yacine El Younoussi$^{(\boxtimes)}$,
and Ferdaouss Achbal$^{(\boxtimes)}$

Information System and Software Engineering, National School of Applied
Sciences, Abdelmalek Essaadi University, Tetouan, Morocco
Chaymae.az@gmail.com, yacine.info@gmail.com,
ferdaouss.achbal@gmail.com

Abstract. This study attempts to describe and discuss the different approaches and methods dedicated to Named Entity Recognition (NER) systems in various languages, in order to justify the choice of a distributional approach for an Arabic NER system using deep learning methods and a Neural Network word representation (Embeddings) as an add-in feature in the unsupervised learning process.

Keywords: Machine learning · NLP · Arabic NER · Deep learning network
Embeddings · CharWNN model

1 Introduction

Since the nascence of terms like "Big data" that eases the ability to harness a big amount of non-filtered information, the need of linguistic and deep semantic analysis has become a necessity to achieve a better and faster decision-making by understanding the meaning of human written texts beyond a simple keyword match. The field of Natural Language Processing (NLP) aims to equip a machine with the human-language-like knowledge to fulfill this understanding task.

Most state-of-the-art NLP applications, such as Statistical Machine Translation [1], Information Extraction [2] and Opinion mining holder recognition [3], rely on Named Entity Recognition (NER). This task aims to identify highlighted entities that hold key information for the language processing, and classifies them into predefined semantic classes [4].

Due to human languages' peculiarities, especially in the semantic field, word sense ambiguities makes creating NER System a challenging task. Which becomes more harder for Arabic Language because of the nature of its morphology, since the general form of a word in Arabic is: Prefix (es) +…+ Stem +…+ Suffix (es), (e.g./'will you remember us'/ translates as أَسَتَتَذَكَّرُونَنَا /'/ أَ ـ سَـتَ ـ تَ ـ ذَكَّرـونَ ـنَا /'/).

In Addition, there is the difficulty of Arabic characteristics such as diacritics that are no longer used in the Modern Standard Arabic (MSA) form, and lack of capitalization, which is very important in NER process of other languages like English.

In order to create a NER system able to overcome these difficulties, the NLP researchers community proceeded with three main approaches: The Rule-based, The ML-based (Machine Learning) and the hybrid approach, from which the ML-based approaches gained more popularity by giving impressive Recall (R) results despite the fact that semantic understanding is still a far distant goal. However, since the human language developed in a way that reflects the innate ability provided by the brain's Neural Network (NN), an Artificial NN word representation seems to be a good method if we want to reach even best results, and semantic disambiguation.

The rest of this paper is structured as follows: Sect. 2 gives a background information on Arabic NER, Sect. 3 illustrates the Deep Neural Network (DNN) architecture, and discusses the choice of a NN Word representation. We conclude by introducing our approach and perspective for future works.

2 Named Entity Recognition Task in Arabic

In computational linguistics, Word Sense disambiguation (WSD) is an open problem of natural language processing. It is basically identifying the most appropriate sense of ambiguous words given their contexts [5].

Named Entity Recognition is a core task to perform the WSD. According to Benajiba and Rosso in [6], the Sixth Message Understanding Conference (MUC-6)[1] defined the NER task as three subtasks: ENAMEX for the proper names (Organizations, Person, and Location), TIMEX for temporal expressions and NUMEX for numeric expression (monetary expressions, percentage…).

NER is a challenging task because of the natural language ambiguities, especially if we are talking about a language known by its morphological complexity like Arabic. In Sect. 2.1, we will focus on Arabic's characteristics and semantic ambiguities, and in Sect. 2.2, we will describe some of the previous NLP works on Arabic NER to overcome these difficulties.

2.1 Arabic Challenges

Arabic characteristics. Arabic is considered one of the highly inflected languages, with a rich morphology and a complex syntax. Despite the effort made in the Arabic NLP, many of its tasks including NER remain challenging due to the following characteristics:

Arabic Scripts. Arabic has 28 different isolated letters, and the majority of these letters have three different shapes, according to its position in the word, i.e. initial (ﻫ), middle (ﻬ), last (ﻪ). Moreover, Arabic's language in use can be classified into three

---

[1] http://cs.nyu.edu/cs/faculty/grishman/muc6.html.

types [7]: Classical Arabic (CA) is the language used in most of the Arabic religious texts (Quran, Sunnah). Modern Standard Arabic (MSA) the language used today in writing. The Colloquial Arabic Dialects which is the language used by Arabs to communicate in their informal daily life, it is a regional variant that differs across regions of the different Arab countries.

Diacritics. The majority of the MSA texts are written without considering the use of diacritics; these diacritics play the role of five main short vowels not included in the alphabet [8]. They may wipe off grammatical and semantic ambiguity, thus, one word can return all its possible morphological variations, which makes it necessary to consider the word's context in order to predict the correct vocalization.

Word structure. Verbs and nouns derive from a root of three letters called stem, it forms the basic building block of the Arabic word. Arabic word-forms are complex units [9]; one word can be structured as Fig. 1 shows, using the stem of the word, proclitic, prefixes, suffixes, and enclitics.

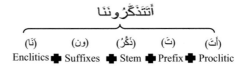

Fig. 1. Arabic word structure

Semantic ambiguity. In addition to Arabic characteristics that makes it one of the most complex morphological languages, Semantic ambiguity, such as polysemy and metonymy, makes Arabic NLP applications an even more difficult goal to achieve.

Polysemy. It can be defined as words with the capacity to have multiple meanings (e.g. Fig. 2). This phenomenon is a common semantic ambiguity between Arabic and other languages. Although, the absence of diacritics in Arabic does affect in the ambiguity degree.

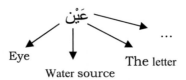

Fig. 2. Example of word polysemy

Metonymy. As Brun et al. defined in [10], metonymy is the use a word attached to an entity to designate another, where the two entities are linked by functional logic relationship.

2.2   Related Works

To overcome the Arabic Language ambiguities in creating NER System, the NLP researchers have proceeded with three main approaches [11]:

Rule-based approach. This approach relies on lexicons and a set of rules, these extraction rules are produced by linguistic experts.

ML-based approach (Machine Learning). This approach aims to learn the rules of entity extraction in an autonomous way. The acquisition of these rules is learned from a large corpus in a supervised, semi-supervised or an unsupervised learning method. The supervised learning approach is when the corpus is already annotated, where for the semi-supervised and the unsupervised learning approach; the manual and the laborious work is reduced because they request a small or non-annotated corpus.

Hybrid approach. As its noun indicates, this approach integrates the rule-based and ML-based approaches to optimize overall performance.

In this section, we will focus on some of the related works of Arabic NER in which they used the ML-based or the hybrid approaches. These researches will be divided according to what model is used:

Support Vector Machines Model. SVM model is a supervised machine-learning algorithm, that can be used for either classification or regression challenges. More specifically, it constructs hyper-plane of training data in a high-or-infinite dimensional space. Bunescu and Paşca's goal in [12] is to detect whether a proper name refers to a named entity included in the dictionary[2], and then disambiguates between multiple NEs that can be denoted by the same proper name. Their approach consists of using a SVM disambiguation kernel to train and exploit the high coverage and rich structure of the knowledge encoded in an online encyclopedia. The results were significantly outperforming, their model shows that using the Wikipedia taxonomy leads to a substantial improvement in accuracy and a less informed baseline.

Maximum Entropy (ME) Model. This Model aims to precious the stated prior data and the probability distribution, thus presenting the current state of knowledge by the largest Entropy[3]. One of the well-known NER System for Arabic is Benajiba and Rosso's work in [13], their approach consists of firstly detecting NE boundaries, in which the authors determine a list of features to be set for the GIS (General Iterative Scaling) algorithm to estimate the deferent weights λ_i. The next step is creating a model to use for combining between the ME-based boundaries detector and the POS-tags-based detector in order to build a classifier. Finally, this classifier computes for each word the probability of the considered classes using a BIO[4] system, e.g. B-Person, I-Person, O, B-Location…

---

[2] Their dictionary was created by using Encyclopedia.

[3] Entropy is a measure of uncertainty.

[4] The BIO method stand for: the Beginning, the inside and the Outside of the Entity.

Conditional Random Field Model. CRF is a sequence Model of statistical modeling method used for structured prediction, thus it takes word's context into account. CRF models are mostly used to encode known relationships between observations and construct consistent interpretations. In order to boost their system, the authors of [13] added the CRF approach to their system, Therefore, ANERsys2.0[5] was created to show impressive efficiency and Recall[6]. The author's goal was the pre-treatment of texts to separate between the prefixes, suffixes, and lemmas. This approach helped in the process of capturing morpho-syntactic characteristics.

3   A Deep Neural Network Approach for NER

The human brain is quite proficient at language processing, therefore word sense disambiguation (WSD). The natural language is formed in a way that requires so much of the brain's reflection of the neurologic reality. In other terms, the human language developed in a way that reflects the innate ability provided by the brain's neural networks.

In the Artificial Intelligence world, it has been a long-term challenge to develop the ability in computers to do machine learning for NLP, until the researchers developed the concept of Deep Neural Network (DNN); a computational simulation to the hierarchical organized hidden layers (neurons) of the brain. The DNN methods are designed to cluster and classify the input data in order to recognize numerical, texts, images or even sound patterns.

In this section, we will describe the DNN approach architecture. Then we will present and explain our NER approach.

3.1   DNN and NLP Applications

Principal of neural network. Classification is the process of categorizing a group of objects while using some basic data features that describes them. Nowadays, we can find several classifiers such as Logistic Regression, Naïve Bayes, SVM that we described briefly in Sect. 2.2, and Neural Networks. The aim of a classifier is producing a score; this score is calculated by processing the received set of data as inputs, the return output is the one with the highest score.

The thing that differentiates a neural network classifier from other classifiers is using multiple highly structured layers as shown in Fig. 3; this feature can be helpful when an output of classifier can fall in more than two different categories, therefore it is resolving the problem of pattern complexity.

---

[5] ANERsys is the name of the Arabic NER system created by BENAJIBA and his team of researchers, it's available at: http://users.dsic.upv.es/ ~ ybenajiba.

[6] Recall is an evaluation measure, used for NLP application.

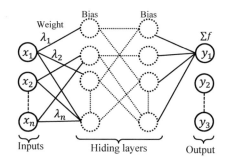

Fig. 3. A deep Neural Network architecture

Each set of input is defined by unique weights λ^i (the edges) and bias w^i (the nodes), which means that the combination used for each activation is also unique; therefore each node has a different output.

Machine learning with a DNN approach. Training is the process that ensures the high accuracy of the net; in this step, the weights and bias are set by changing their values after each iteration. To train the net, each generated output is compared to the actual correct output; the point here is to make the difference between the two as small as possible across millions of training examples, by twisting the weights and bias values. As a result of a well-trained Neural Network, the generated outputs are accurate at each time.

3.2 DNN Approach for NER

General architecture for NLP tasks. The concept of a DNN approach is to extract a set of automatic features from the input sentence using several layers of feature extraction, to be then fed to a classification algorithm. This features are automatically trained according to the desired NLP task [14]. A Neural Network (NN) automatically learns features in the deep layers of its architecture.

A DNN approach for NER. As we stated in previous sections, the WSD is still a far goal to reach, despite the effort made by Arabic NLP researchers in boosting NER Systems' performances.

Our study is based on Guimaraes and dos Santos approach in [15] for a neural character Embeddings for a Spanish and Portuguese NER System, the authors proposed an independent NER system in which, unlike most NER systems that rely on the outputs of other NLP applications such as POS-tagging and chunking. This approach aims to use only automatically learned features. It is based on the principle of using Embeddings; word-level representation to learn semantic features, and character-level representation to learn morphological features, in order to perform sequential classification (Fig. 4).

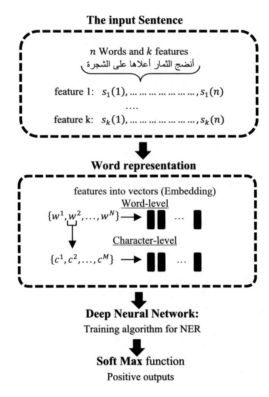

Fig. 4. DNN approach for NER

As far as we know, there is still no work on deep learning approaches for Arabic NER using word Embeddings.

Word Embeddings. The word Embeddings is a neural language model that captures morphological, syntactic and semantic information about words. This technique enables efficient computation of words similarities, through low dimension matrix operation, which will be useful in treating Arabic characteristics and semantic ambiguity problems. The NLP field proposed various deep learning methods to learn word vector representations.

Word2vec[7]. [16] has been one of the most used tool. It consists mostly of the Continuous Bag of Words (CBOW) model to predict a word given its context, and the Skip-Gram (SG) model to predict the surroundings of a given word. We are more interested in the SG model, since we aim to perform pre-training of Word-level embeddings using the SG Neural Network architecture.

---

[7] A tool created by Mikolov, it's a group of related models used to create word Embeddings <https://github.com/dav/word2vec>.

CharWNN Model. This word representation employs a convolutional layer that allows effective character-level feature extraction from words of any size. As the authors of [17] have explained, given a sentence, the network gives for each word a score for each class $\tau \in T$, using a BIO system:

$$T = \left\{O, B_{Person}, I_{Person}, B_{Organization}, I_{Organization}, \dots\right\}$$

Where B_{person}, I_{Person} present respectively the beginning and the Inside of an entity person, and O is the outside of this entity.

In order to score a word, the network takes as input a fixed-sized window of words centralized in the target word as illustrated in Fig. 5. The input is passed through a sequence of layers where features with increasing levels of complexity are extracted. The first layer of the network is used to transform words into real-valued feature vectors.

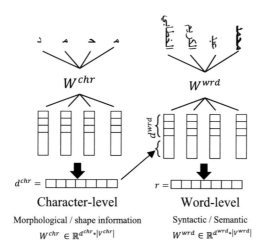

Fig. 5. CharWNN embeddings

According to Guimaraes and dos Santos in [15], a fixed-sized word vocabulary V^{wrd} is needed, in which the words are composed of characters from a fixed-sized character vocabulary V^{chr}. Thus, a sentence composed of N words $\left\{w^1, w^2, \dots, w^N\right\}$, whose w^i. is converted into a vector $U_n = \left[r^{wrd}; r^{chr}\right]$, which is composed of two sub-vectors: the word-level embedding $r^{wrd} \in \mathbb{R}^{d^{wrd}}$ that capture syntactic and semantic information and the character-level embedding $r^{wch} \in \mathbb{R}^{d^{chr}}$ that capture morphological and shape information.

Word level Embeddings. Are basically encoded by column vectors in an embedding matrix $W^{wrd} \in \mathbb{R}^{d^{wrd} * |V^{wrd}|}$, where the matrix W^{wrd} is a parameter to be learned, and the size of the word-level embedding d^{wrd} is a hyperparameter to be set by us. Word-level Embeddings are meant to capture syntactic and semantic information, which are crucial to the NER task.

Character-Level Embeddings. They are computed for each word by convolutional layer, given a word w composed of M characters $\{c^1, c^2, \ldots, c^M\}$, the goal is to transform each character c^M into a character embedding r_m^{chr}. Logically to form an embedded matrix, Character Embeddings are also encoded by column vectors in the embedding matrix $W^{chr} \in \mathbb{R}^{d^{chr}*|V^{chr}|}$. Given a character c, its embedding r^{chr} should be obtained by the matrix-vector product: $r^{chr} = W^{chr} v^c$, where v^c is a vector of size $|V^{chr}|$ that has the value 1 at index c and zero in all other positions.

Scoring the Word vectors. The model used in [15] uses a prediction scheme that considers the sentence structure. The idea is the use of a transition score A_{t_u} for jumping from tag $t \in T$ to $u \in T$ in successive words, and a score A_{0t} for starting from the tth tag. Given a sentence $[w]_1^N = \{w_1, \ldots, w_N\}$ the score for tag path $[t]_1^N = \{t_1, \ldots, t_N\}$ is computed as follow:

$$S([w]_1^N, [t]_1^N, \theta) = \sum_{n=1}^{N} \left(A_{t_{n-1} t_n} + s(w_n)_{t_n} \right) \tag{1}$$

θ is the set of all trainable network parameters where:

$$\begin{cases} W^1 \in \mathbb{R}^{hl_u \times k^{wrd}(d^{wrd} + cl_u)} \text{ and } W^2 \in \mathbb{R}^{|T| \times hl_u} \\ b^1 \in \mathbb{R}^{hl_u} \text{ and } b^2 \in \mathbb{R}^T \end{cases}$$

$s(w_n)_{t_n}$ is the score given for the tag t_n at word w_n where:

$$s(w_n) = W^2 h(W^1 r + b^1) + b^2 \tag{2}$$

This last score is computed after extracting the level of representation by the neural network layers that proceed the vector r resulting from the concatenation of a sequence of k^{wrd} embeddings centralized in the nth word, while:

$$r = \left(u_n - \frac{(k^{wrd} - 1)}{2}, \ldots, u_n + \frac{(k^{wrd} - 1)}{2} \right)^T \tag{3}$$

Network training. To interpret the sentence score in (1) as a conditional probability over a path, the authors of [15] proposed to exponentiate the score (1) and normalize it. By taking the log, the conditional log-probability will be defined as:

$$\log p([t]_1^N | [w]_1^N, \theta) = S([w]_1^N, [t]_1^N, \theta) - \log \left(\sum_{\forall [u]_1^N \in T^N} e^{S([w]_1^N, [t]_1^N, \theta)} \right) \tag{4}$$

The process of training CharWNN word Embeddings is done by computing the log-likelihood equation in (4) using dynamic programming; minimizing the negative log-likelihood using SGD[8], and computing the gradient of the NN relying on the BPN[9].

Recent work has shown that large improvements in terms of model accuracy can be obtained by performing unsupervised pre-training of word Embeddings. Thus, we intend to perform an unsupervised learning method using tools like NLTK[10], GATE[11], and Word2vec.

Embeddings for Arabic. The first step that proceeds a word Embeddings for MSA, is preparing the corpus by crawling it using for example SpiderLing[12] to create a large enough corpus for unsupervised training. Then comes the Data cleaning step we can use JusText[13] tool to exclude non-text data such as navigation links and advertisement, and Onion[14] tool recommended by Dahou et al. in [18] to detect and remove duplicated data. Afterwards, it comes the actual Word Embedding step, in order to capture syntactic and semantic word relationships from very large data sets. We can use the Skip-Gram model from the Word2Vec tool for CharWNN, to benefit from its fast training and good result.

4   Conclusion

In this study, we described a DNN approach for the NER task that jointly uses word-level and character-level representation to boost the performance of Arabic NER systems. The idea behind this approach is training without any other NLP applications' outputs or handcrafted features. We conclude that the same DNN approach applied for Spanish/Portuguese NER system in [15] can also achieve outstanding results for Arabic NER, by overcoming Arabic morphological difficulties.

We hope these insights will facilitate further research into improving the Arabic NER task, and possibly achieve the word sense disambiguation.

References

1. Do, Q.K.: Apprentissage Discriminant des Modèles Continus en Traduction Automatique. Université Paris-Saclay (2016)
2. Algahtani, S.: Arabic Named Entity Recognition: A Corpus-Based Study. University of Manchester (2011)
3. Elarnaoty, M., AbdelRahman, S., Fahmy, A.: A machine learning approach for opinion holder extraction in Arabic language. Am. Control Conf. **3**(2), 4479–4484 (2015)

---

[8] Stochastic Gradient Descent code <https://github.com/mateuszmalinowski/SGD>.

[9] Back-Propagation Net <https://backpropagation-neural-network.soft112.com/>.

[10] Natural Language toolkit is a leading platform for building Python programs to work with human language data <http://www.nltk.org>.

[11] An integrated development environment for text engineering <https://gate.ac.uk/>.

[12] crawler for effective creation and annotation of linguistic corpora < http://corpus.tools/browser/spiderling>.

[13] A heuristic based boilerplate removal tool <http://code.google.com/p/justext/>.

[14] A de-duplication tool <https://code.google.com/p/onion>.

4. Darwish, K.: Information retrieval. In: Hirst, G., Hovy, E., Johnson, M. (eds.) Natural Language Processing of Semitic Languages, pp. 299–334. Springer, Berlin (2014)
5. Yepes, A.J.: Word embeddings and recurrent neural networks based on Long-Short Term Memory nodes in supervised biomedical word sense disambiguation. J. Biomed. Inform. **73**, 137–147 (2016)
6. Benajiba, Y., Rosso, P.: ANERsys 2.0: conquering the NER task for the Arabic language by combining the maximum entropy with POS-tag information. In: 3rd Indian International Conference on Artificial Intelligence, pp. 1814–1823 (2007)
7. Shaalan, K.: A survey of Arabic named entity recognition and classification. Comput. Linguist. **40**(September 2012), 469–510 (2013)
8. Yacine, E.Y.: Towards an Arabic web-based information retrieval system (ARABIRS): stemming to indexing. Int. J. Comput. Appl. **109**(14), 16–21 (2015)
9. Fyshe, A., Murphy, B., Talukdar, P., Mitchell, T.: Supervised morphological segmentation in a low-resource learning setting using conditional random fields (2013)
10. Brun, C., Ehrmann, M., Jacquet, G.: Résolution de métonymie des entités nommées: proposition d'une méthode hybride. TAL **50**, 87–110 (2009)
11. Abdallah, Z.S., Carman, M., Haffari, G.: Multi-domain evaluation framework for named entity recognition tools. Comput. Speech Lang. **43**, 34–55 (2017)
12. Bunescu, R., Paşca, M.: Using encyclopedic knowledge for named entity disambiguation. In: Proceedings of the 11th Conference of the European Chapter of the Association for Computational Linguistics, pp. 9–16, April 2006
13. Benajiba, Y., Rosso, P., BenedíRuiz, J.M.: ANERsys: an Arabic named entity recognition system based on maximum entropy. In: CICLing 2007, pp. 143–153 (2007)
14. Levy, O., Goldberg, Y.: Dependency-based word embeddings. In: Proceedings of the 52nd Annual Meeting of the Association for Computational Linguistics, vol. 2, pp. 302–308 (2014)
15. Guimaraes, V., dos Santos, C.N.: Boosting named entity recognition with neural character embeddings. In: Proceedings of the Fifth Named Entity Workshop, Joint with 53rd ACL and the 7th IJCNLP, pp. 25–33 (2015)
16. Levy, O., Goldberg, Y.: Neural word embedding as implicit matrix factorization. In: Advances in Neural Information Processing Systems, pp. 2177–2185 (2014)
17. Zadrozny, B., dos Santos, C.N.: Learning character-level representations for part-of-speech tagging. In: Proceedings of the 31st International Conference Machine Learning, vol. 32 (2014)
18. Dahou, A., Xiong, S., Zhou, J., Haddoud, M.H.: Word embeddings and convolutional neural network for Arabic sentiment classification. In: Proceedings of the 26th International Conference on Computational Linguistics, pp. 2418–2427 (2016)

Discrete Particle Swarm Optimization for Travelling Salesman Problems: New Combinatorial Operators

Morad Bouzidi[1(✉)], Mohammed Essaid Riffi[1], and Ahmed Serhir[2]

[1] LAROSERI Laboratory, Department of Computer Science,
Chouaib Doukkali University, El Jadida, Morocco
mrbouzidi@gmail.com
[2] Department of Mathematics, Chouaib Doukkali University,
El Jadida, Morocco

Abstract. The Particle Swarm Optimization is one of the most famous nature inspired algorithm that belongs to the swarm optimization family. It has already been used successfully in the continuous problem. However, this algorithm has not been explored enough for the discrete domain. In this work we introduce new operators that are dedicated to combinatorial research that we implemented on a modified discrete particle swarm optimization called DPSO-CO to solve travelling salesman problem. The experimental results on a set of different instances, and the comparison study with others adaptations show that adopting new ways, combinations and operators gives birth to a really competitive efficient algorithm in operational research.

Keywords: Combinatorial research · Discrete operators
Particle swarm optimization · Travelling salesman problem

1 Introduction

The metaheuristic algorithms, is a set of nature inspired methods, generally based on search probability technique to find a good solution within a reasonable time. Today, the metaheuristic algorithms have become more involved into solving a real engineer single and multi-objective problems; before applying the metaheuristics on a real problem, the researchers tested its quality performance to resolve classical problems. As far as the combinatorial problems is concerned, the travelling salesman problem (TSP) is considered as the most famous one that belongs to NP-hard problems [1], Its resolution consists in finding a short travel among a set of cities that the salesman has to visit in order to finish his journey at the initial starting point. Furthermore, TSP has several variations in different areas, which makes its resolution more attractive in computer wiring, vehicle routing and scheduling problems. Moreover, numerous metaheuristics have been published for TSP such as genetic algorithm [2], harmony search algorithm [3], particle swarm optimization [4], ant colony optimization [5] and bee colony optimization [6].

© Springer International Publishing AG, part of Springer Nature 2018
A. Abraham et al. (Eds.): SoCPaR 2017, AISC 737, pp. 141–150, 2018.
https://doi.org/10.1007/978-3-319-76357-6_14

The particle swarm optimization (PSO) [7] is one of the well known optimization algorithms, inspired by bird flocks intelligence method, and classified on swarm intelligence algorithm. The principle of PSO focuses on the collaboration of all particles swarm, where each particle seeks a good position and moves with a specific velocity. This algorithm defines a specific rule which allows each particle to follow the best position in swarm without ignoring its own search making it always search for what is best. In fact, several adaptations of PSO called discrete particle swarm optimization (DPSO) have been proposed [8–10] for a different combinatorial optimization problems, one of the most known adaptations was introduced by Clerc [11], but the experience showed that adaptation of DPSO does not give competitive results like other metaheuristics methods.

This work presents a new adaptation of DPSO named DPSO-CO with the introduction of new combinatorial operators that follows the idea of Clerc in addition to unification dimension, which is then concluded by the presentation of a performance DPSO. The rest of this paper is organized as follows: definition of travelling salesman problem in the Sect. 2 and description of particle swarm optimization algorithm in the Sect. 3, while the Sect. 4 is devoted to give presentation of new operators and a new adaptation. The application and the study of experimental results of this proposition comes in the Sect. 5, and the conclusion and perspective in the Sect. 6.

2   Travelling Salesman Problem

The travelling salesman problem consists on finding the shortest path for the salesman to take in order to visit all cities while two rules must be respected: first, he needs to visit each city only once, second, he has to finish at the starting point. Mathematically speaking, the solution of this problem is a permutation π, contains n elements, where n presents the number of cities, and the ith element π_i indicate the ith city that should be visited, therefore, the objective function to be minimized is:

$$f(x) = \sum_{i=1}^{n-1} distance(\pi_i, \pi_{i+1}) + distance(\pi_n, \pi_1) \qquad (1)$$

3   Particle Swarm Optimization

In 1995, Kennedy and Eberhart introduced the particle swarm optimization [12] that is a research algorithm based on the cooperation and sharing information within a defined research space in order to find a good food source like the social behaviour in bird flocking and fish schooling for example In researcher space, the algorithm launches a set of individuals as candidate solution called "particles", each particle has a position known by all the group. In every moment, particles are moving with a specific velocity towards the best position without ignoring their previous ones that will be shared in the group once found.

More concretely, in D-dimensional space, a population of n particles as potential solutions, each particle has a position P^t (Eq. 2) which is generated randomly in $t = 0$.

$$P^t = \left(X_i^t\right)_{i=1,...,n} = \left(x_{i1}^t, x_{i2}^t, ..., x_{iD}^t\right)_{i=1,...,n} \tag{2}$$

However, before next iteration t, each i th particle is based on the global best position founded in group $gbest^{t-1} = \left(gbest_k^{t-1}\right)_{k=1,...,D}$, and its best previous position $pbest_i^{t-1} = \left(pbest_{i,k}^{t-1}\right)_{k=1,...,D}$ to calculate its next velocity $V_i^t = \left(v_{i,k}^t\right)_{k=1,...,D}$ according Eq. 3, which will take it towards another position $X_i^t = \left(x_{i,k}^t\right)_{k=1,...,D}$ by formulation of Eq. 4.

$$V_i^t = \omega \times V_i^{t-1} + c_1 \times r_1 \times \left(pbest_i^{t-1} - X_i^{t-1}\right) + c_2 \times r_2 \times \left(gbest^{t-1} - X_i^{t-1}\right) \tag{3}$$

$$X_i^t = X_i^{t-1} + V_i^t \tag{4}$$

Where ω is the inertia coefficient constant in interval $[0, 1]$, c1 and c2 are cognitive and social parameters constants in interval $[0, 2]$, and r_1, r_2 are two generated parameters randomly in $[0, 1]$. This process keeps repeating until the stopping condition will be reached. As in our case, the objective is to find the best cycle of TSP without exceeding a maximum iteration number, Algorithm 1 resumes this process with a pseudo code of PSO.

Algorithm 1. Pseudo code of proposed method

```
Begin
Initial population P with d solutions randomly
For each p in P do
   Initial their velocity : vp
   Initial their best position founded  : pbest
end For each
Get gbest : the best pbest in P
While ((optimum not finder) and (iteration number is not
archived)) do
   For each p in P do
      Calculate vp with equation 3
      Update p following equation 4
      if(cost(p)<cost(pbest)) then
         pbest = p
         if(cost(p)<cost(gbest)) then
            gbest = p
         end if
      end if
   end for each
end while
return gbest
End.
```
---

4 Proposed Particle Swarm Optimization with Discrete Operators

The proposed discrete particle swarm optimization with combinatorial operators (DPSO-CO), follows the same method of operators presented by Clerc [11], but they focus mainly on the unification of the dimension of velocity and position, in order to avoid any truncate of velocity so that no data would be lost. This section is devised in three sub-sections, where the first one contains a presentation of the novel operators while in the second one adaptation of DPSO for this operator is to be found.

4.1 Novel Discrete Operators

Before starting representing operators, as it is known, the structure of the position is represented as a permutation π, and velocity is a set of permutation. In this proposed operators, we defined the data structure of velocity like position, which means that the velocity is a permutation of d elements, where d is dimension of research space.

$$v = (v_k)_{k=1,...,d} \tag{5}$$

Addition for position and velocity. The result of addition between velocity and position is a position, which velocity translates the order of item position, that means in each ith item of x will be the $(v_i)^{th}$ item in x' resumed in Eq. 6, to clarify this operation formulate 7 and Fig. 1. Example of position x plus velocity vshow example of addition between position and velocity.

$$x + v = x' = \left(x_i'\right)_{i=1,...,d} = (x_k)_{i=1,...,d}/i = v_k \tag{6}$$

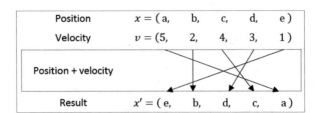

Fig. 1. Example of position x plus velocity v

$$\left. \begin{array}{l} x = (1,4,3,5,2) \\ v = (5,2,4,3,1) \end{array} \right\} x + v = (2,4,5,3,1) \tag{7}$$

With this operator, if addition of position and two velocities give the same position, it implies that they are equal (Eq. 8), and this redefinition makes more sense of vector translation to velocity.

$$x + v' = x + v'' \Rightarrow v' = v'' \tag{8}$$

Addition between two velocities. The addition of velocity v^1 with another velocity v^2 is a new velocity v^3 with d elements, where each item i in v^3 equal value of $(v_i^1)^{th}$ item in v^2, this action resumes to translation of v^1 by v^2, where $(x + v^1) + v^2 = x + (v^1 + v^2)$ but $v^1 + v^2$ not equals $v^2 + v^1$. Equation 9 resumes an example of this operation.

$$v^1 + v^2 = v^3 = \left(v_i^3 \right)_{i=1,...,d} = \left(v_{(v_i^1)}^2 \right)_{i=1,...,d} \tag{9}$$

$$\left. \begin{array}{l} v^1 = (2,5,1,4,3) \\ v^2 = (3,1,2,5,4) \end{array} \right\} v^3 = v^1 + v^2 = (1,4,3,5,2) \tag{10}$$

Subtraction operator. At the same idea of addition operator, it is applied between two positions, so the x^1 minus x^2 is v where each ith item in v (v_i^{th}) equal the rang k flowing Eq. 11, the result is a velocity v which enables the second position to move towards the first position. And if $x^1 - x^2 = v$ then $x^1 = x^2 + v$.

$$x^1 - x^2 = v = (v_i)_{i=1,...,d} = (k)_{i=1,...,d} / x_k^1 = x_i^2 \tag{11}$$

$$\left. \begin{array}{l} x^1 = (3,1,4,2,5) \\ x^2 = (1,4,3,5,2) \end{array} \right\} v = x^1 - x^2 = (2,3,1,5,4) \tag{12}$$

Multiplication operator. In this operator, the coefficient represents a probability parameter of adjustment for velocity, in other way, the particle moves with some velocity which can include some problems that can create a small disruptive impact on its position for the next iteration. Thus, the multiplication between a coefficient c and velocity v. After that the operator checks if a random number between 0 and 1 is less than c, then he applies a random swap in v, otherwise it does nothing (Eq. 13).

$$cv = \begin{cases} random\ swap\ of\ v & rand() < c \\ v & otherwise \end{cases} \tag{13}$$

4.2   Discrete Particle Swarm Optimization

The proposal method DPSO-CO based on these new defined operators generates randomly particles population, then starts the search loop, where it detects, in each iteration, g_{best} and generates the next solution for each particle following Eqs. 3 and 4, where the inertia coefficient constant takes value 1, c1 and c2 are generated for each iteration between 0 and 1. Each new solution gets a small perturbation with 2-opt in order to look for a better neighbour, 2-opt improves a solution by applying iteratively exchanges between two edges resuming in Fig. 2. 2-Opt Swap, where (A) represents the initial case, and (B) is the new perturbation of tour. In TSP case, to accept exchange

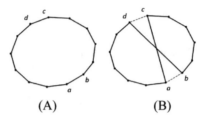

(A)   (B)

Fig. 2. 2-Opt Swap

between two pairs (a, b) and (c, d) to (a, c) and (b, d) the sum of distances of (a, c) and (b, d) must be less than the sum of pairs (a, b) and (d, c). This process continues until DPSO-CO finds the best known solution for the problem or if it exceeds the maximal number of possible iterations.

5   Experimental Results

In order to validate its performances, the proposed adaptation has been programmed with C++ language and has been made on PC with processor Intel(R) Core(TM) i5-2500 CPU @ 3.30 GHz and 4 Go of RAM, and has been tested on some symmetric benchmarks of TSPLIB library [13].

Table 1 represents the numerical results on twenty instances of symmetric problems, each instance is executed thirty times, the first column represents the name of instance problem, the second column Opt shows the best known optimum of instance, while the third and fourth columns show respectively the best and worst solutions obtained by DPSO-CO. In the fifth column the average length of all solutions is to be found, and in the next column the standard deviation SD of all the results is shown. The seventh and eighth columns represent respectively the percentage relative error of the average PDav and the best solution PDbest according to the optimal known solution presented in second column, this value is calculated following Eq. 14. Finally in the ninth column $C_{1\%}/C_{opt}$, where C1% indicates the number of solutions where relative error is less than 1 and Copt is the number of solutions equal to optimum known solution that means the number of iteration which its relative error is null, and the last one is time column that shows the average of time execution of all iterations in second.

$$PDsolution = \frac{Solution\ lenght - optimal\ known\ lenght}{optimal\ known\ lenght} \times 100\% \qquad (14)$$

The experimental results have shown that this proposed method gets the optimum one for the majority of instances, especially for medium-sized category instances where the average execution time is the shortest. To verify its performance according to the current DPSO, we have implemented DPSO presented by Clerc [11] and showed the results in Table 2. Comparison of the proposed algorithm with DPSO proposed by Clerc [11]. When comparing, DPSO gets the best known solutions for some instances, but in general, it does not provide results better than DPSO-CO that give solutions in a

Table 1. Results of the proposed method DPSO-CO for some symmetric instance of TSPLib.

Instance	Opt	Best	Worst	Average	SD	PDav	PDbest	$C_{1\%}/$ C_{opt}	Time
eil51	426	**426**	427	426.07	0.25	0.02	**0.00**	30/28	0.22
berlin52	7,542	**7,542**	**7,542**	**7,542.00**	**0.00**	**0.00**	**0.00**	**30/30**	0.02
st70	675	**675**	**675**	**675.00**	**0.00**	**0.00**	**0.00**	**30/30**	0.13
eil76	538	**538**	540	538.53	0.72	0.01	**0.00**	30/18	2.09
pr76	108,159	**108,159**	**108,159**	**108,159.00**	**0.00**	**0.00**	**0.00**	**30/30**	0.11
rat99	1,211	**1,211**	1,212	1,211.07	0.25	0.01	**0.00**	30/28	2.29
kroA100	21,282	**21,282**	**21,282**	**21,282.00**	**0.00**	**0.00**	**0.00**	**30/30**	0.28
kroB100	22,141	**22,141**	**22,141**	**22,141.00**	**0.00**	**0.00**	**0.00**	**30/30**	1.63
kroD100	21,294	**21,294**	21,309	21,294.50	2.69	**0.00**	**0.00**	30/29	2.25
kroE100	22,068	**22,068**	22,106	22,069.40	6.85	0.01	**0.00**	30/28	3.93
rd100	7,910	**7,910**	7,911	79,10.07	0.25	**0.00**	**0.00**	30/28	2.19
eil101	629	**629**	635	631.20	1.78	0.35	**0.00**	30/06	6.49
lin105	14,379	**14,379**	**14,379**	**14,379.00**	**0.00**	**0.00**	**0.00**	**30/30**	0.27
pr107	44,303	**44,303**	44,387	44,309.50	18.43	0.01	**0.00**	30/26	5.48
ch130	6,110	**6,110**	6,147	61,22.57	11.44	0.21	**0.00**	30/08	15.79
pr136	96,772	**96,772**	97,105	96,870.70	86.98	1.10	**0.00**	30/07	18.27
pr144	58,537	**58,537**	**58,537**	**58,537.00**	**0.00**	**0.00**	**0.00**	**30/30**	0.48
Ch150	6,528	**6,528**	6,561	6,537.37	10.91	0.14	**0.00**	30/16	20.45
kroA150	26,524	**26,524**	26,689	26,556.20	41.66	0,12	**0.00**	30/03	27.18
kroB150	26,130	**26,130**	26,237	26,152.70	23.61	0,09	**0.00**	30/04	28.01
rat195	2,323	2,336	2,370	2,355.70	7.20	1,41	0.56	02/00	65.60
kroA200	29,368	**29,368**	29,599	29,495.30	55.63	0,43	**0.00**	30/01	83.45
ts225	126,643	**126,643**	**126,643**	**126,643.00**	0,00	0,00	**0.00**	**30/30**	0.86
tsp225	3,916	3,939	3,983	3,964.90	12.18	1.25	0.59	02/00	64.67

shorter time and with less amount of relative errors terms of who get result in small time and with a less reduced amount of relative errors. Both Fig. 3. Comparison average relative error between proposed method and PSO-ACO-3Opt [14] and Fig. 4. Comparison average execution time between proposed method and PSO-ACO-3Opt [14] show the difference of the PDav and the execution time between the proposed DPSO and the latest improved hybrid DPSO with Ant colony and 3-opt called PSO-ACO-3opt [14], from these figures DPSO-CO is more efficient than PSO-ACO-3Opt in terms of time and quality.

In addition, Table 3. Comparison of the proposed algorithm with DCS [15]. compares results of DPSO-CO with DCS [15], where column Time represents the average time for all iteration average execution, and the last line represents the average of each column. The results show that their performances are very close, and DPSO-CO gives a nice result especially in time of execution for medium-size instances, but when the size of the problem increases, DCS takes the advantage.

148   M. Bouzidi et al.

Table 2. Comparison of the proposed algorithm with DPSO proposed by Clerc [11].

Instance	DPSO				DPSO-CO			
	PDbest	PDav	$C_{1\%}/C_{Opt}$	Time	PDbest	PDav	$C_{1\%}/C_{Opt}$	Time
eil51	**0.00**	0.14	**30**/13	2.12	**0.00**	0.02	**30**/28	0.22
st70	**0.00**	0.01	**30**/29	4.82	**0.00**	**0.00**	**30/30**	0.13
eil76	**0.00**	0.94	15/03	7.87	**0.00**	0.01	**30**/18	2.09
pr76	**0.00**	0.04	**30**/15	8.38	**0.00**	**0.00**	**30/30**	0.11
kroB100	**0.00**	0.17	**30**/06	21.57	**0.00**	**0.00**	**30/30**	1.63
kroD100	**0.00**	0.38	**30**/02	21.18	**0.00**	**0.00**	**30**/29	2.25
kroE100	0.13	0.40	**30**/00	21.84	**0.00**	0.01	**30**/28	3.93
eil101	0.79	1.84	02/00	20.80	**0.00**	0.35	**30**/06	6.49
lin105	**0.00**	**0.00**	**30/30**	20.89	**0.00**	**0.00**	**30/30**	0.27
pr107	**0.00**	0.21	**30**/01	22.81	**0.00**	0.01	**30**/26	5.48
ch130	0.20	0.90	19/00	51.93	**0.00**	0.21	**30**/08	15.79
pr136	0.21	0.76	26/0	44.14	**0.00**	1.10	**30**/07	18.27
pr144	**0.00**	**0.00**	**30/30**	31.79	**0.00**	**0.00**	**30/30**	0.48
Ch150	0.34	1.08	12/00	64.49	**0.00**	0.14	**30**/16	20.45
kroA150	0.36	1.04	12/00	71.22	**0.00**	0.12	**30**/03	27.18
kroB150	0.24	0.86	24/00	68.39	**0.00**	0.09	**30**/04	28.01
rat195	2.28	3.37	00/00	145.70	0.56	1.41	02/00	65.60
kroA200	0.85	1.31	05/00	184.59	**0.00**	0.43	**30**/01	83.45
ts225	0.07	0.31	30/0	216.03	**0.00**	**0.00**	**30/30**	0.86

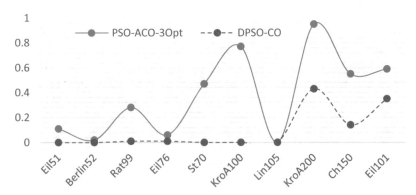

Fig. 3. Comparison average relative error between proposed method and PSO-ACO-3Opt [14].

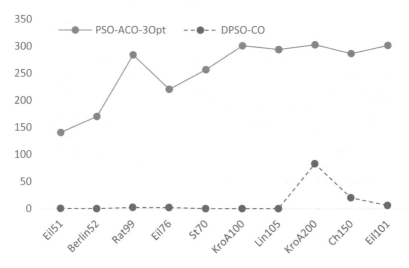

Fig. 4. Comparison average execution time between proposed method and PSO-ACO-3Opt [14].

Table 3. Comparison of the proposed algorithm with DCS [15].

Instance	PDbest		PDav		$C_{1\%}/C_{Opt}$		Time	
	DCS	DPSO-CO	DCS	DPSO-CO	DCS	DPSO-CO	DCS	DPSO-CO
eil51	**0.00**	**0.00**	**0.00**	0.02	**30/30**	30/28	1.16	0.22
st70	**0.00**	**0.00**	**0.00**	**0.00**	**30/30**	**30/30**	1.56	0.13
eil76	**0.00**	**0.00**	**0.00**	0.01	30/29	30/18	6.54	2.09
pr76	**0.00**	**0.00**	**0.00**	**0.00**	**30/30**	**30/30**	4.73	0.11
kroB100	**0.00**	**0.00**	**0.00**	**0.00**	30/29	30/30	8.74	1.63
kroD100	**0.00**	**0.00**	0.04	**0.00**	30/19	30/29	8.74	2.25
kroE100	**0.00**	**0.00**	**0.00**	0.01	30/18	30/28	14.18	3.93
eil101	**0.00**	**0.00**	0.22	0.35	30/06	30/06	18.74	6.49
lin105	**0.00**	**0.00**	**0.00**	**0.00**	**30/30**	**30/30**	5.01	0.27
pr107	**0.00**	**0.00**	**0.00**	0.01	30/27	30/26	12.89	5.48
ch130	**0.00**	**0.00**	0.42	0.21	28/07	30/08	23.12	15.79
pr136	0.01	**0.00**	0.24	1.10	30/00	30/07	35.82	18.27
pr144	**0.00**	**0.00**	**0.00**	**0.00**	**30/30**	**30/30**	2.96	0.48
Ch150	**0.00**	**0.00**	**0.33**	0.14	29/10	30/16	27.74	20.45
kroA150	**0.00**	**0.00**	0.17	0.12	30/07	30/03	31.23	27.18
kroB150	**0.00**	**0.00**	0.11	0.09	30/05	30/04	33.01	28.01
rat195	0.04	0.56	0.81	1.41	20/00	02/00	57.25	65.60
kroA200	0.04	**0.00**	0.26	0.43	29/00	30/01	62.08	83.45
ts225	**0.00**	**0.00**	0.01	**0.00**	30/26	30/30	47.51	0.86
Average	0.01	0.01	0.14	0.21	29/18	29/19	21.21	14.89

6   Conclusion

This paper presents a new adaptation of DPSO-CO characterized by a novel definition of discrete operators. This proposed DPSO-CO method has been applied on different symmetric TSP instances from TSPLib. The result was compared in confrontation with the last hybrid PSO known and a recent competitive algorithm DCS. From this obtained study, it can be concluded that the proposed DPSO-CO are effective and more performant than other methods. However, this work opens new horizons for DPSO-CO to solve other combinatorial problems especially when using open discrete operators in new different ways that can be used in the future for others metaheuristic algorithms to test and increase the performance of research.

References

1. Arora, S.: Polynomial time approximation schemes for Euclidean traveling salesman and other geometric problems. J. ACM (JACM) **45**, 753–782 (1998)
2. Grefenstette, J., Gopal, R., Rosmaita, B., Van Gucht, D.: Genetic algorithms for the traveling salesman problem. In: Proceedings of the First International Conference on Genetic Algorithms and their Applications. Lawrence Erlbaum, Hillsdale (1985)
3. Bouzidi, M., Riffi, M.E.: Adaptation of the harmony search algorithm to solve the travelling salesman problem. J. Theor. Appl. Inf. Technol. **62**(1) (2014)
4. Wang, K.P., Huang, L., Zhou, C.G., Pang, W.: Particle swarm optimization for traveling salesman problem. In: 2003 International Conference on Machine Learning and Cybernetics (2003)
5. Dorigo, M., Birattari, M.: Ant colony optimization. In: Encyclopedia of Machine Learning. Springer (2010)
6. Wong, L.P., Low, M.Y.H., Chong, C.S.: A bee colony optimization algorithm for traveling salesman problem. In: Modeling & Simulation, AICMS 2008 (2008)
7. Kennedy, J.: Particle swarm optimization. In: Encyclopedia of Machine Learning, pp. 760-766. Springer, US (2011)
8. Chen, A.L., Yang, G.K., Wu, Z.M.: Hybrid discrete particle swarm optimization algorithm for capacitated vehicle routing problem. J. Zhejiang Univ.-Sci. A **7**(4), 607–614 (2006)
9. Kennedy, J., Eberhart, R.C.: A discrete binary version of the particle swarm algorithm. In: Systems, Man, and Cybernetics, 1997. Computational Cybernetics and Simulation (1997)
10. Pan, Q.K., Tasgetiren, M.F., Liang, Y.C.: A discrete particle swarm optimization algorithm for the no-wait flowshop scheduling problem. Comput. Oper. Res. **35**(9), 2807–2839 (2008)
11. Clerc, M.: Discrete particle swarm optimization, illustrated by the traveling salesman problem. In: New Optimization Techniques in Engineering, pp. 219–239 (2004)
12. Eberhart, R., Kennedy, J.: A new optimizer using particle swarm theory. In: Proceedings of the Sixth International Symposium on Micro Machine and Human Science (1995)
13. Reinelt, G.: TSPLIB—A traveling salesman problem library. ORSA J. Comput. **3**(4), 376–384 (1991)
14. Mahi, M., Baykan, Ö.K., Kodaz, H.: A new hybrid method based on particle swarm optimization, ant colony optimization and 3-opt algorithms for traveling salesman problem. Appl. Soft Comput. **30**, 484–490 (2015)
15. Ouaarab, A., Ahiod, B., Yang, X.S.: Discrete cuckoo search algorithm for the travelling salesman problem. Neural Comput. Appl. **24**(7–8), 1659–1669 (2014)

On Maximal Frequent Itemsets Enumeration

Said Jabbour[1(✉)], Fatima Zahra Mana[1,2], and Lakhdar Sais[1]

[1] CRIL-CNRS, Université d'Artois, 62307 Lens Cedex, France
{jabbour,sais}@cril.fr, fatimana93@gmail.com
[2] INPT, Institut national des postes et télécommunications, Rabat, Morocco

Abstract. Enumerating interesting patterns from data is an important data mining task. Among the set of possible relevant patterns, maximal frequent patterns is a well known condensed representation that limits at least to some extent the size of the output. Recently, a new declarative mining framework based on constraint programming (CP and satisfiability (SAT) has been designed to deal with several pattern mining tasks. For instance, the itemset mining problem has been modeled as a constraint network/propositional formula whose models correspond to the pattern to be mined. In this framework, closeness, maximality and frequency properties can be handled by additional constraints/formulas. In this paper, we propose a new propositional satisfiability based approach for mining maximal frequent itemsets that extends the one proposed in [13]. We show that instead of adding constraints to the initial SAT based itemset mining encoding, the maximal itemsets, can be obtained by performing clause learning during search. Our approach leads to a more compact encoding. Experimental results on several datasets, show the feasibility of our approach.

1 Introduction

Mining frequent itemsets in datasets is a fundamental problem in the data mining field since it enables several mining tasks such as discovering association rules, data correlations, sequential patterns, etc. The problem of finding frequent itemsets and its corresponding association rules was originally proposed by Agrawal in [2]. Several algorithms have been proposed to deal with this enumeration problem. One can cite the levelwise and the pattern-growth like approaches. The first one is based on the generate-and-test framework [2], while the second is based on the divide-and-conquer framework [9] (see also H-Mine [18], LCM [22] and [21] for a survey).

Usually, the number of the frequent patterns is known to be large. To reduce the size of the output, condensed representations, such as maximal and closed itemsets, have been introduced. Maximal Frequent Itemsets (MFI) is a subset of closed itemsets with longest size. Several methods have been proposed to discover the maximal frequent itemsets. Bayardo proposed a MaxMiner algorithm which extends the Apriori algorithm [15]. MaxMiner employs a breadth-first traversal of the search space to limit the database scanning. Furthermore,

© Springer International Publishing AG, part of Springer Nature 2018
A. Abraham et al. (Eds.): SoCPaR 2017, AISC 737, pp. 151–160, 2018.
https://doi.org/10.1007/978-3-319-76357-6_15

it uses a dynamic heuristic to increase the effectiveness of superset-frequency pruning. Several other enhancements have been suggested for mining the MFI. Pincer-Search algorithm combined the top-down and bottom-up [16] techniques to discover the maximal frequent itemset. Agarwal et al. implement a depth first search technique with bitmap representation (DepthProject) [1]. In which, column denotes the items and rows denotes the transactions. Like MaxMiner algorithm, they used dynamic reordering and look-ahead pruning. A projection mechanism is used to reduces the size of database. They efficiently find the support counts and give a superset of the MFI. D. Burdick et al. further extended DepthProject and named it as Mafia [4]. They used vertical bit-vector data format. Compression and projection on bitmaps are applied to increase the performance. Unlike DepthProject and MaxMiner pruning technique, Mafia used Parent Equivalence Pruning. Algorithm GenMax [8] is a backtrack search based algorithm. More specifically, it integrates numerous optimization techniques to prune the search space including progressive focusing that perform maximality checking and diffset propagation for fast support counting. To search MFI, SmartMiner [24] records at each step tail information to guide the search for new MFI.

Recently, constraint programming and propositional satisfiability have been used for modeling several data mining tasks in a declarative way, and solving them using generic solving techniques [5,7,11,13,19]. Such approaches show that many task in data mining such as itemset, association rules, and sequence mining can be reduced into the enumeration of the models of a set of constraints/propositional formulas. In [20], the authors show that the generation of MFI can also be formulated as the enumeration of a set of models of a constraint network by adding a constraint to force the required models to be maximal.

In this paper, we propose to extend the SAT based approach for mining closed frequent itemsets proposed in [13], to generate the set of maximal ones. We show that instead of encoding maximality as a set of constraints, we can add clauses during search to cut non-maximal frequent itemsets.

2   Background

Let us first introduce the propositional satisfiability problem (SAT) and some necessary notations. We consider the conjunctive normal form (CNF) representation for the propositional formulas. A *CNF formula* Φ is a conjunction ($\wedge$) of clauses, where a *clause* is a disjunction ($\vee$) of literals. A *literal* is a positive (p) or negated ($\neg p$) propositional variable. The two literals p and $\neg p$ are called *complementary*. A CNF formula can also be seen as a set of clauses, and a clause as a set of literals. We denote by $Var(\Phi)$ the set of propositional variables occurring in Φ.

A *Boolean interpretation* $\mathcal{B}$ of a propositional formula Φ is a function which associates a value $\mathcal{B}(p) \in \{0,1\}$ (0 corresponds to *false* and 1 to *true*) to the propositional variables $p \in Var(\Phi)$. It is extended to CNF formulas as usual. A *model* of a formula Φ is a Boolean interpretation $\mathcal{B}$ that satisfies the formula, i.e., $\mathcal{B}(\Phi) = 1$. We note $\mathcal{M}(\Phi)$ the set of models of Φ. *SAT problem* consists in deciding if a given formula admits a model or not.

Algorithm 1. DPLL Based Enumeration solver

Input: Φ: a CNF formula
Output: $\mathcal{M}$: all the models of Φ

```
 1  I = ∅, ;                                          /* Current interpretation */
 2  M = ∅ ;                                                    /* Set of models */
 3  dl = 0 ;                                                   /* decision level */
 4  while (true) do
 5      conflictClause = unitPropagation(Φ, I);
 6      if (conflictClause!=null) then
 7          if (dl == 0) then return M;
 8          ℓ = lastDecision;
 9          backtrack(dl − 1);
10          dl = dl − 1;
11          I = I ∪ {¬ℓ};
12      else
13          if (I ⊨ Φ) then
14              M = M ∪ {I};
15              ℓ = lastDecision;
16              backtrack(dl-1);
17              dl = dl − 1;
18              I = I ∪ {¬ℓ};
19          else
20              ℓ = selectDecisionVariable(Φ);
21              dl = dl + 1;
22              I = I ∪ {ℓ};
23          end
24      end
25  end
```

Let us informally describe the most important components of modern SAT solvers, usually called CDCL (Conflict Driven Clause Learning) solvers. They are based on a reincarnation of the historical Davis, Putnam, Logemann and Loveland procedure, commonly called DPLL [6]. It performs a backtrack search; selecting at each level of the search tree, a decision variable which is set to a Boolean value. This assignment is followed by an inference step that deduces and propagates some forced unit literal assignments. This is recorded in the implication graph, a central data-structure, which encodes the decision literals together with there implications. This branching process is repeated until finding a model or a conflict. In the first case, the formula is answered satisfiable, and the model is reported, whereas in the second case, a conflict clause (called learnt clause) is generated by resolution following a bottom-up traversal of the implication graph [17,23]. The learning or conflict analysis process stops when a conflict clause containing only one literal from the current decision level is generated. Such a conflict clause asserts that the unique literal with the current level (called asserting literal) is implied at a previous level, called assertion level, identified as the maximum level of the other literals of the clause. The solver backtracks to the assertion level and assigns that asserting literal to *true*. When an empty conflict clause is generated, the literal is implied at level 0, and the original formula can be reported unsatisfiable. In addition to this basic scheme, modern SAT solvers use other components such as activity based heuristics and restart policies. An extensive overview can be found in [3].

When tackling the enumeration of the models of a CNF formula, the clause learning components can be heavy when the set of models to learn is very large. In [10], the authors show that a simple DPLL algorithm can outperform CDCL based approach when dealing with the problem of enumerating models of propositional formulas. The efficiency of the proposed enumeration algorithm have been evaluated on several propositional formulas encoding itemsets mining problem [14].

Algorithm 1 depicts the general scheme of a DPLL-like procedure to enumerate the models of a formula. when a model is found (line 16), then the solver performs a simple backtracking to the precedent level to propagate the negation of the last decision (line 18). The enumeration stops when a conflict occurs in level 0 or if the model is found at level 0.

3   Frequent Itemset Mining

A database D comprises a set of transactions $\{T_1, T_2, \ldots, T_m\}$ and Ω a set of items. Each transaction has a unique transaction identifier (tid) and contains a set of items over Ω. A set of items is often called an itemset. Let I be an itemset and T a transaction. We will use the notation, $I \subseteq T$, to denote that I is a subset of the set of items that T contains. When the context is clear, we will often directly refer to a transaction as the set of items that it contains.

Classical Data mining problems are usually concerned with itemsets that frequently occur in a database of transactions. The number of occurrences of an itemset in a database is commonly referred to as the support of this itemset, formalized as follows.

Definition 1 (Support). *Let X be an itemset and D a database of transactions. The support of X in D, denoted $Supp(X)$, is the number of transactions of D in which X occurs as a subset. The frequency of X is defined as the ratio of $Supp(X)$ to $|D|$.*

Let $\mathcal{D}$ be a transaction database over Ω and λ a minimum support threshold. The *frequent itemset mining problem* consists in computing the following set:

$$FI(D, \lambda) = \{X \subseteq \Omega \mid Supp(X) \geqslant \lambda\}.$$

Mining of the complete set of frequent itemsets may lead to a huge number of itemsets. In order to face this issue, this problem can be limited on the extraction of closed itemsets. Thus, enumerating all closed itemsets allows us to reduce the size of the output.

Table 1. A transaction database D

tid	Itemset			
1	A	B	C	D
2	A	B	E	F
3	A	B	C	
4	A	C	D	F
5	G			
6	D			
7	D	G		

Definition 2 (Closed Frequent Itemset). *Let D be a transaction database (over Ω). An itemset X is a closed itemset if there exists no itemset X' such that (1) $X \subseteq X'$ and (2) $\forall T \in D, X \in T \rightarrow X' \in T$.*

Extracting all the elements of $FI(\mathcal{D}, \lambda)$ can be obtained from the closed itemsets by computing their subsets. We denote by $CFI(\mathcal{D}, \lambda)$ the subset of all closed itemsets in $FI(\mathcal{D}, \lambda)$.

For instance, consider the transaction database described in Table 1. The set of closed frequent itemsets with the minimal support threshold equal to 2 are: $CFI(\mathcal{D}, 2) = \{A, D, G, AB, AC, AF, ABC, ACD\}$.

If we consider subset inclusion as defining a partial order for itemsets, then we can introduce the notions of maximal frequent itemsets, as follows.

Definition 3 (Maximal Frequent Itemset). *Let X be an itemset of a database D. We say that X is a maximal itemset in D given a minimum threshold λ, if there exists no itemset Y such that $X \subset Y$ and Y is a frequent in D.*

The problem of mining maximal frequent itemsets MFI, is to enumerate all maximal frequent itemsets whose support is no less than a preset threshold.

$$MFI(D, \lambda) = \{X \subseteq \Omega \mid Supp(X) \geqslant \lambda \text{ and } X \text{ is maximal}\}.$$

4 Itemset Mining Based Constraints Encoding

Recent work tackle the problem of mining frequent itemsets by encoding this problem into constraints. In [13], the authors show that the generation of itemsets from transaction D can be encoded as the enumeration of the models of a CNF formula Φ i.e., there exists a one to one mapping between the models of the set of constraints and the set of frequent itemsets. Let us first review this approach. As said, the approach of [13] consists to introduce boolean variables p_a (resp. q_i) to represent each item $a \in \Omega$ (resp. each transaction T_i).

$$\bigwedge_{i=1}^{m} (\neg q_i \leftrightarrow \bigvee_{a \in \Omega \setminus T_i} p_a) \tag{1}$$

$$\sum_{i=1}^{m} q_i \geqslant \lambda \qquad (2)$$

$$\bigwedge_{a \in \Omega} ((\bigvee_{a \notin T_i} q_i) \vee p_a) \qquad (3)$$

The formula (1) allows to model the transaction database and then to catch the itemsets when an itemset appear in a transaction T_i ($q_i = 1$) if and only iff the variables not involved in T_i are set to false. The formula ($\neg q_i \leftrightarrow \bigvee_{a \in \Omega \setminus T_i} p_a$) can be translated to the following CNF formula:

$$\bigwedge_{a \in \Omega \setminus T_i} (\neg q_i \vee \neg p_a) \wedge (q_i \vee \bigvee_{a \in \Omega \setminus T_i} p_a)$$

Formula (2) allows us to consider the itemsets having a support greater than or equal to λ. This encoding is defined as a 0/1 linear inequality, usually called cardinality constraint. Because of the presence of such constraint in several applications, many efficient CNF encodings have been proposed over the years. Mostly, such encodings try to derive the best compact representation while preserving the efficiency of constraint propagation (e.g. [12]).

Formula (3) capture the closure property. Intuitively, if the itemset is involved in all transactions containing an item a then a must be added to the candidate itemset. In other words, when in all the transactions where a does not appear, the candidate itemset is not included, we deduce that the candidate itemset appears only in transactions containing the item a. Consequently, to be closed, the item a must be added to the final itemset.

The advantage of this approach is its ability to be easy to modify in order to integrate others constraints. For instance, enumerating itemsets of size at most k, can be expressed by simply adding the linear constraint $\sum_{a \in \Omega} p_a \leqslant k$.

5   Enumerating Maximal Itemset

In this section we show how to enumerate maximal itemset using constraints. Let us recall that in [20], the authors provide a set of constraints that can be added to the CSP formula of closed frequent itemsets problem in order to generate the set of maximal itemset. Similarly, using satisfiability the maximality can be expressed as follow:

$$\bigwedge_{a \in \Omega} (\neg p_a \rightarrow \sum_{T_i \mid a \in T_i} q_i < \lambda) \qquad (4)$$

Constraint (4) expesses that if a is not in the final maximal itemsets I, it means that the frequency of I in the transactions containing a is lower than λ. However, translating constraint (4) into clauses can lead to a large CNF formula. Another alternative is to manage theses constraints inside the solver. However their number remains a problem for the efficiency of SAT solvers. To avoid constraint addition, in this section we show a new approach that allows to add blocking clauses during search. For this purpose, we consider that our solver is a DPLL-like procedure. To illustrate our approach, assume that the solver assign with the truth value true the variables representing items. Let us note $\mathcal{B}$ as the model of the CNF formula encoding the frequent itemsets mining task and $P(\mathcal{B}) = \{a|\mathcal{B}(p_a) = 1\}$ the frequent itemset. Clearly the first found model $\mathcal{B}$ corresponds to a maximal frequent itemset $P(\mathcal{B})$. To avoid retrieving a model $\mathcal{B}'$ such that $P(\mathcal{B}') \subset P(\mathcal{B})$, one need to block all itemset $I \subset \mathcal{B}$. To do this, it is sufficient to add the blocking clause $c = (\bigvee_{a \in \Omega \setminus P(\mathcal{B})} p_a)$ to Φ. The solver can then backtrack and performs positive assignment of the variables representing Ω.

The main idea of our approach consists in adding blocking clauses each time a model is found. Let us note that such clauses comprise the literals representing items that are assigned to false under the current assignment. Consequently, such clauses are false before backtracking. In order to enumerate correctly the set of maximal itemsets, one need to take into account the levels of literals of c to backtrack at the adequate level.

Note that in general for real transaction databases, the missing items in each transaction T_i is larger than those of T_i i.e., $|T_i| \ll |\Omega \setminus T_i|$. As a consequence, each blocking clause c added to cut non maximal itemsets can be larger. Let us suppose that the current itemset appears in the transaction T_i. Clearly, c can be written as $c = (\bigvee_{a \in T_i \setminus P(\mathcal{B})} p_a \vee \bigvee_{a \in \Omega \setminus T_i} p_a)$. On the other hand, according to constraints (1), $\neg q_i = \bigvee_{a \in \Omega \setminus T_i} p_a$. So, c can be rewritten as $c = (\bigvee_{a \in T_i \setminus P(\mathcal{B})} p_a \vee \neg q_i)$. As formulated, the size of c can be considerably reduced. Furthermore, one can obtain the smallest clause c by choosing the smallest transaction T_i containing $P(\mathcal{B})$.

In the sequel, we present Algorithm 2 that can be integrated to Algorithm 1 to generate maximal itemsets based on the approach discussed below. More precisely, and as discussed before, when a model is found, a blocking clause c must be added to the formula. This clause is falsified at the current level. Furthermore, the literals of c can be assigned at different level of the search tree. Then, instead of performing a simple backtracking as shown in Algorithm 1, we have to analyze c in order to deduce the adequate backtracking level.

Algorithm 2. DPLL for Maximal Itemsets

---

1 $\mathcal{M} = \mathcal{M} \cup \{\mathcal{B}\}$;
2 $c \to \bigvee_{a \in \Omega \ | \ \neg p_a \in \mathcal{B}} p_a$;
3 $\Phi \leftarrow \Phi \cup c$;
4 $btl, \ell \leftarrow analyze(c)$;
5 backtrackUntil(btl);
6 $dl = btl$;
7 $\mathcal{B} = \mathcal{B} \cup \{\ell\}$;

Example 1. Let us reconsider the set of transactions of Table 1 and assume that $\lambda = 2$. Suppose that the solver choose the following variable ordering p_A, p_E, p_F, p_G, p_B, p_D, and p_C. Then, the first assignment leading to a model is $\mathcal{B} = \{p_A, \neg p_E, \neg p_F, \neg p_G, p_B, \neg p_D, p_C\}$ such that p_A is assigned to level 1, p_B at level 2, and p_C at level 3. The added blocking clause is $c = (p_D \vee p_E \vee p_F \vee p_G)$. Then the solver must backtrack to the level 1 since c is falsified in level 2.

From the complexity point of view, as shown, each time a model is found a blocking clause is added to the formula. Consequently, the maximum number of added clauses is equal to the number of maximal itemsets. Let us recall that the number of maximal itemsets is often limited compared to closed ones.

6   Experimental Validation

We carried out an experimental evaluation to analyze the effect of adding blocking clauses and branching heuristics. To this end, we implemented a DPLL-like procedure, denoted `DPLL4MFI` as described in Algorithm 2. In this procedure, each time a model is found, we add a blocking clause (no-good) and perform a backtracking after analyzing the blocking clause. We considered a variety of datasets taken from the `FIMI`[1] and `CP4IM`[2] repositories. All the experiments were done on Intel Xeon quad-core machines with 32 GB of RAM running at 2.66 GHz. For each instance, we used a timeout of 15 min of CPU time.

In our experiments, we compare the performances of `DPLL4MFI` to the variant `DPLL4CFI` that enumerate the set of closed itemsets. Let us recall that the DPLL procedure utilize a static branching heuristic i.e., the variables corresponding to the infrequent items are assigned first. Note that the solver does not branch on q_i variables. In fact, by assigning the variables p_a, $a \in \Omega$, the variables q_i, $1 \leq i \leq m$ are propagated because the variables q_i depend on those of p_a as expressed by constraint (1).

In Table 2 we report some obtained results when considering some datas and some chosen values of the minimum support threshold. Unsurprisingly, the

Table 2. Maximal Itemsets for some datas

instance (#item, #trans)	min_supp	#CFI	#MFI	DPLL4CFI time(s)	DPLL4MFI time(s)
hepatitis (68, 137)	14	1827263	189205	2.21	29.35
lumph (68, 148)	10	46801	5191	0.08	0.11
primary_tumor (31, 336)	34	31024	2043	0.04	0.02
anneal (93, 812)	81	1224754	15977	2.7	0.47
german-credit (112, 1000)	100	2080152	232107	10.38	37.26
mushroom (119, 8124)	812	3287	453	1.95	1.72

---

[1] FIMI: http://fimi.ua.ac.be/data/.
[2] CP4IM: http://dtai.cs.kuleuven.be/CP4IM/datasets/.

number of maximal itemsets is often limited compared to closed ones. Also, the time needed to compute them is slightly greater than the one needed to generate the closed ones due to added blocking clauses.

7   Conclusion

In this paper, we consider the problem of enumerating the set of maximal frequent itemsets using propositional satisfiability. We show how to answer this question by utilizing a DPLL-like procedure for model enumeration combined to clause learning from models. Experimental results show that this approach is interesting, as it allows to avoid adding maximality constraints to the initial encoding.

As a future work, we plan to pursue our investigation in order to find the best heuristics to speed up the enumeration of the models. For example, it would be interesting to integrate some background knowledge in such heuristic design. Finally, clause learning, an important component for the efficiency of SAT solvers, admits several limitation in the context of model enumeration. An important issue, is to study how such important mechanism can be efficiently integrated when maximal itemset generation is considered.

References

1. Agarwal, R.C., Aggarwal, C.C., Prasad, V.V.V.: Depth first generation of long patterns. In: Proceedings of the Sixth ACM SIGKDD International Conference on Knowledge Discovery and Data Mining, KDD 2000, pp. 108–118 (2000)
2. Agrawal, R., Imieliński, T., Swami, A.: Mining association rules between sets of items in large databases. In: Proceedings of the 1993 ACM SIGMOD International Conference on Management of Data, SIGMOD 1993, pp. 207–216. ACM, New York (1993)
3. Biere, A., Heule, M.J.H., van Maaren, H., Walsh, T. (eds.): Handbook of Satisfiability. Frontiers in AI and Applications, vol. 185. IOS Press, Amsterdam (2009)
4. Burdick, D., Calimlim, M., Gehrke, J.: MAFIA: a maximal frequent itemset algorithm for transactional databases. In: ICDE, pp. 443–452 (2001)
5. Coquery, E., Jabbour, S., Saïs, L., Salhi, Y.: A sat-based approach for discovering frequent, closed and maximal patterns in a sequence. In: Proceedings of the 20th European Conference on Artificial Intelligence (ECAI 2012), pp. 258–263 (2012)
6. Davis, M., Logemann, G., Loveland, D.W.: A machine program for theorem-proving. Commun. ACM **5**(7), 394–397 (1962)
7. Gebser, M., Guyet, T., Quiniou, R., Romero, J., Schaub, T.: Knowledge-based sequence mining with ASP. In: Proceedings of the Twenty-Fifth International Joint Conference on Artificial Intelligence, IJCAI 2016, New York, NY, USA, 9–15 July 2016 (2016)
8. Gouda, K., Zaki, M.J.: Efficiently mining maximal frequent itemsets. In: Proceedings of the 2001 IEEE International Conference on Data Mining, San Jose, California, USA, 29 November–2 December 2001, pp. 163–170 (2001)
9. Han, J., Pei, J., Yin, Y.: Mining frequent patterns without candidate generation. SIGMOD Rec. **29**, 1–12 (2000)

10. Jabbour, S., Lonlac, J., Sais, L., Salhi, Y.: Extending modern SAT solvers for models enumeration. In: Proceedings of the 15th IEEE International Conference on Information Reuse and Integration, IRI 2014, Redwood City, CA, USA, 13–15 August 2014, pp. 803–810 (2014)
11. Jabbour, S., Sais, L., Salhi, Y.: Boolean satisfiability for sequence mining. In: 22nd ACM International Conference on Information and Knowledge Management (CIKM 2013), pp. 649–658. ACM (2013)
12. Jabbour, S., Sais, L., Salhi, Y.: A pigeon-hole based encoding of cardinality constraints. TPLP **13**(4–5-Online-Suppl.) (2013)
13. Jabbour, S., Sais, L., Salhi, Y.: The top-k frequent closed itemset mining using top-k SAT problem. In: European Conference on Machine Learning and Knowledge Discovery in Databases (ECML/PKDD 2003), pp. 403–418 (2013)
14. Jabbour, S., Sais, L., Salhi, Y.: On SAT models enumeration in itemset mining. CoRR abs/1506.02561 (2015)
15. Bayardo Jr., R.J.: Efficiently mining long patterns from databases. In: SIGMOD 1998, Proceedings ACM SIGMOD International Conference on Management of Data, Seattle, Washington, USA, 2–4 June 1998, pp. 85–93 (1998)
16. Lin, D.I., Kedem, Z.M.: Pincer-search: a new algorithm for discovering the maximum frequent set, pp. 103–119 (1998)
17. Marques-Silva, J.P., Sakallah, K.A.: GRASP - a new search algorithm for satisfiability. In: Proceedings of IEEE/ACM CAD, pp. 220–227 (1996)
18. Pei, J., Han, J., Lu, H., Nishio, S., Tang, S., Yang, D.: H-mine: hyper-structure mining of frequent patterns in large databases. In: Proceedings IEEE International Conference on Data Mining, ICDM 2001, pp. 441–448 (2001)
19. Raedt, L.D., Guns, T., Nijssen, S.: Constraint programming for itemset mining. In: ACM SIGKDD, pp. 204–212 (2008)
20. Raedt, L.D., Guns, T., Nijssen, S.: Constraint programming for itemset mining. In: Proceedings of the 14th ACM SIGKDD International Conference on Knowledge Discovery and Data Mining, Las Vegas, Nevada, USA, 24–27 August 2008, pp. 204–212 (2008)
21. Tiwari, A., Gupta, R., Agrawal, D.: A survey on frequent pattern mining: current status and challenging issues. Inf. Technol. J. **9**, 1278–1293 (2010)
22. Uno, T., Kiyomi, M., Arimura, H.: LCM ver. 2: efficient mining algorithms for frequent/closed/maximal itemsets. In: FIMI 2004, Proceedings of the IEEE ICDM Workshop on Frequent Itemset Mining Implementations, Brighton, UK, 1 November 2004 (2004)
23. Zhang, L., Madigan, C.F., Moskewicz, M.W., Malik, S.: Efficient conflict driven learning in Boolean satisfiability solver. In: IEEE/ACM CAD 2001, pp. 279–285 (2001)
24. Zou, Q., Chu, W.W., Lu, B.: SmartMiner: a depth first algorithm guided by tail information for mining maximal frequent itemsets. In: Proceedings of the 2002 IEEE International Conference on Data Mining (ICDM 2002), Maebashi City, Japan, 9–12 December 2002, pp. 570–577 (2002)

Particle Swarm Optimization as a New Measure of Machine Translation Efficiency

José Angel Montes Olguín[1(✉)], Jolanta Mizera-Pietraszko[2],
Ricardo Rodriguez Jorge[3], and Edgar Alonso Martínez García[3]

[1] Instituto Tecnológico Superior Zacatecas Norte,
Río Grande, Zacatecas, Mexico
anxelm@gmail.com
[2] Opole University, pl. Kopernika 11a, 45-040 Opole, Poland
[3] Universidad Autonoma de Ciudad Juarez, Av. Del Charro #450 norte,
Partido Romero, Ciudad Juarez, 32310 Chihuahua, Mexico

Abstract. The present work proposes a new approach to measuring efficiency of evolutionary algorithm-based Machine Translation. We implement some attributes of evolutionary algorithms performing cosine similarity objective function of a Particle Swarm Optimization (PSO) algorithm then, we evaluate an English text set for translation precision into the Spanish text as a simulated benchmark, and explore the backward process. Our results show that PSO algorithm can be used for translation of multiple language sentences with one identifier only, in other words the technology presented is language-pair independent. Specifically, we indicate that our cosine similarity objective function improves the velocity attribute of the PSO algorithm, making the complex cost functions unnecessary.

Keywords: Evolutionary algorithms · Machine Translation · Cosine similarity

1 Introduction

Machine Translation (MT) refers to computerized systems that are responsible for producing translations with, or without human assistance. The first recorded attempts to machine-aided translation were reported in the seventeenth century, when mechanical dictionaries were build with the aim at overcoming the language barriers. In the past decades, many technologies have been applied to make automatic translation comparable to professional human interpretation. At first, researchers realized that translation process turned out to be much more complex than assumed before [1]. The first expert systems relied on heuristic and statistical methods until the most recent neural network technologies have been initiated. Currently, neural network machine translation is a real breakthrough when the users all over the world have noticed the significant improvement in Google Translation services since in 2016 Google decided to train NN on utterances entered to their search engines in order to collocate the words in as many as possible contexts. Many other companies started to develop such emerging

© Springer International Publishing AG, part of Springer Nature 2018
A. Abraham et al. (Eds.): SoCPaR 2017, AISC 737, pp. 161–170, 2018.
https://doi.org/10.1007/978-3-319-76357-6_16

technologies as neural machine translation systems and they refined statistical framework. Still, translation quality depends upon the language-pairs, in particular those the most popular amongst the Google users.

Evolutionary algorithms have demonstrated their potential predominantly in intuitive solving optimization problems. Based on the model of natural selection, these algorithms generate a target population, making use of very limited resources. They compete with each other to find the closest problem solution [2]. Some results of the tasks related to MT and Evolutionary Algorithms (EAs) were presented in literature, for example, translate a number of sentences in a specific technical context, automatically train translation models for any language pair and resolve a word sense disambiguation problem [3–5].

On the other side, cosine similarity (CS) functions measure concordance of two sentences in different languages thus we can avoid any ambiguity in sentence translation in the cases when the same words can have a number of different meanings depending on their context. For this reason, the words in the sentence constitute the only indicator to be stored. Velocity of the particle considerably improves the MT efficacy in the translation process. Cosine similarity has been used to measure the semantic similarity and to reduce the error rate by 13% in our experiment. It was also used in an unsupervised speaker adaptation technique, which resulted in better speaker recognition evaluation precision [6, 7].

For this study, the English and Spanish languages are used for the experiment. We merge Particle Swarm Optimization (PSO) and CS at the stage of finding the best solution. To our best knowledge, our approach is novel from the perspective that uses evolutionary algorithms for measuring machine translation efficiency. The PSO algorithm generates the best possible translation when it is given an English sentence to be translated into Spanish and backward. Cosine similarity function comprises the core of the selection process in the PSO system of the most similar candidate to the reference translation.

This paper is organized as follows: In Sect. 2, the methodology is described, we discuss PSP and CS in detail, along with the evolutionary algorithms. Section 3, shows the results obtained after applying the PSO and CS to our methodology as the optimum solution. Finally, Sect. 4 presents the conclusions and our plans for the future work.

2   Methodology

In this research, we address an approach to employ evolutionary algorithms supported with the cosine similarity objective function.

2.1   Evolutionary Algorithms (EA)

First, we start with a very brief description of some EAs. Following [2] the main concepts include:

- The contemporary terminology denotes the whole field by evolutionary computing, the algorithms involved are termed evolutionary algorithms, and it considers evolutionary programming, evolution strategies, genetic algorithms, and genetic programming as subareas belonging to the corresponding algorithm variants.
- The common idea behind all these techniques is the same: given a population of individuals within an environment that has limited resources, competition for those resources causes natural selection. This is done by applying recombination and/or mutation to them.
- EAs are in the family of generate-and-test methods because

  (1) they are population-based,
  (2) most EAs use recombination and
  (3) EAs are stochastic.

- The most important components of EAs are as follows:

  (1) representation,
  (2) evaluation function,
  (3) population,
  (4) parent selection mechanism,
  (5) variation operators, recombination and mutation, and
  (6) survivor selection mechanism.

Some of the works on EAs and machine translation are described below:

[3] presents evolutionary hybrid algorithms to translate sentences in a specific technical context in order to demonstrate that an approach to translation supported with a statistical MT framework, is both feasible and comparable in quality to other techniques that use the same framework, even despite of the other kinds of translators. Their results suggest that implementation of decoder that adapts EDA to some features of a specific statistical model is a good option.

Another study, [4] proposes an evolutionary algorithm for decoding a machine translation process. This method is grounded on the optimization of a total solution. Their performance obtained for the French-English pair were not better than MOSES in terms of bilingual evaluation understudy (BLEU) however, it was comparable with some other works in the field when considered from the perspective of Translation Error Rate (TER). MOSES is a statistical machine translation system that allows one to automatically train translation models for any language pair [8].

The use of EAs is also described by [5], who proposes an interesting approach to word-sense disambiguation by applying genetic and mimetic algorithms. The performance of several models of his algorithms was analyzed based on the experiments carried out on a large Arabic corpus, comparing the results with those using a Naïve Bayes classifier. Finally, he showed that genetic algorithms can achieve more precise predictions than the mimetic algorithms and the Naïve Bayes classifier, achieving up to 79% precision.

2.2   Cosine Similarity

The subsection presents some works that implement cosine similarity to machine translation.

In their research work, [7] they introduce some modification to cosine similarity that does not require explicit score normalization. It relies on a simple mean and covariance statistics gathered from a collection of an impostor speaker *ivectors* to avoid complications related to $z-$ and $t-norm$. This approach further allows for the application of a new unsupervised speaker adaptation technique to the models defined in the *ivector* space. They achieved a *MinDCF* value of 0.0107 on the female English part of the NIST 2008 speaker recognition evaluation.

A method for measuring semantic similarity of texts by using corpus and knowledge-based similarity measures is presented by [6]. In their work, they focus on measuring the semantic similarity of short texts only, and they demonstrate that the semantic similarity method outperforms some other methods based on simple lexical matching, that result in a reduction of the error rate by up to 13% with respect to the traditional vector-based similarity metric. In addition, they show that incorporating semantic information into similarity measures of text increases the likelihood of recognition significantly over both the random baseline and the vector-based cosine similarity baseline, as measured in a paraphrase recognition task. They also show that in some cases, almost all the semantic similarity measures fail; then, the simpler cosine similarity performs better.

2.3   Particle Swarm Optimization Architecture

Research oriented toward statistical natural language processing (NLP) and EAs was conducted by [9], who reviews the best suited problems or specific aspects in those problems aimed to solve them by AE. She concludes that, then the function to optimize is too complex or requires expensive preprocessing to be applied to machine translation, or to parsing complete sentences all alone. She also notes that for some cases when real-time processing is not a demand, a number of NLP problems is not critical; some examples of such cases include corpus alignment for machine translation and grammar induction. The following Algorithm 1 shows the translator function.

Algorithm 1: Translator

Input: The translation mode and the sentence identifier
Output: The target sentence.
1: **Initialization:** Target sentence, PSO problem definition and PSO parameters
2: LanguageLoader (TranslationMode,SentenceIdentifier)
3: PSOTranslator (TranslationMode,SentenceIdentifier)
3: **return** TranslatedSentence

As presented, the translator function calls two external functions in order to process the translation phase. First, the full properties for the target sentence in the language is loaded for translation and then, the PSO objective function is preformed to process the translation phase and to select the best translation for the given sentence.

Algorithm 2: Language Loader

Input: The translation mode and the sentence identifier.
Output: The target language.
1: **Initialization:** The language properties
2: **switch** TranslationMode
3:   **case** 1
4:      TargetLanguage ← LanguageEnglishProperties
5:   **case** n
6:      TargetLanguage ← LanguagenProperties
7: **end switch**

The *LanguageLoader* function presented in Algorithm 2, defines the language into which the source sentence will be uploaded; it is defined as an indicator set that contains the translation and the metadata to improve the process. In Fig. 1, we also show the variables that will be used to run the main PSO tasks: *nPop* for the population, *nvar* for the number of solutions, *BestPersonalPosition* for the best position of a particle, *BestPersonalCost* for the best particle cost found, *Min* and *Max* Velocity for updating the search velocity, *BestSentence* for saving the best solution, *BestSentenceCost* for saving the best cost found, *BestWord* for saving the best word found, and *BestWordCost* for saving and comparing the cost with the search result.

Fig. 1. *LanguageLoader* function and PSO Initialization.

Next, we show some example indicators from a given sentence presented in Table 1:

Table 1. Example of corresponding sentences.

Language	English sentence content (SentenceContent)	Sentence identifier (SentenceId)	Sentence size (SentenceSize)
English	According to the US model	1	5
Spanish	De acuerdo al modelo de Estados Unidos	1	7

In order to translate this sentence, we first define some indicators to identify the main parts and their corresponding translations. This is shown in Table 2.

Table 2. English encoded word.

English Word	Word identifier (WordIdentifier)	Sentence identifier (SentenceId)	Word alignment (WordAlignment)	Word code (WordCode)
According	1	1	1	111

Once we have the sentence indicators identified, we generate the word code. This code is composed of the English word, the word ID to distinguish it from the others, the sentence ID to identify its correspondent sentence, and the word alignment to know the word position in the sentence. Having these indicators defined, we generate the word code.

Every sentence has a corresponding sentence in the target languages. In Table 3, an example of sentence code components are shown.

Table 3. Sentence code components.

	Position 1	Position 2	Position 3	Position 4	Position 5
Word	According	to	the	US	model
Word code	111	212	313	414	515

The encoded sentence from Table 3 is $SentenceCode = 111212313414515$. In Algorithm 3, we show an example of how a sentence is encoded, as well as the encoding of the same sentence in some other languages.

Algorithm 3: PSO Translator

Input: The translation mode and the sentence identifier
Output: The translated sentence.
1: **Initialization:** Target sentence, PSO problem definition and PSO parameters
2: **for** ($i \leftarrow$ 1; nPop)
3:   **for** ($j \leftarrow$ 1; nVar)
4:     **for** ($k \leftarrow$ 1; TargetSentenceSize)
5:       **if** (not BestWordIdStatus)
6:        GeneratedWord $\leftarrow$ random(Word)
7:        **if** (TargetWord = GeneratedWord)
8:          BestWord $\leftarrow$ GeneratedWord
9:        **else**
10:         Similarity $\leftarrow$ CosineSimilarity (TargetWord,GeneratedWord)
11:           **if** (Similarity > BestSimilarity)
12:             BestParticlePosition $\leftarrow$ ParticlePosition
13:             BestParticleCost $\leftarrow$ ParticleCost
14:           **else**
15:             MinVelocity $\leftarrow$ MinVelocity + 1
16:             MaxVelocity $\leftarrow$ MaxVelocity - 1
17:           **end if**
18:         **end if**
19:       **end if**
20:       GeneratedSentence $\leftarrow$ GeneratedSentence + GeneratedWord
21:       **if** (TargetSentence = GeneratedSentence)
22:         BestSentence $\leftarrow$ GeneratedSentence
23:         return GeneratedSentence
24:         break
25:       **end if**
26:       SentenceSim $\leftarrow$ CosineSim (GeneratedSentence, TargetSentence)
27:       **if** (SentenceSim > BestSentenceCost)
28:         BestSentence $\leftarrow$ GeneratedSentence
29:       **end if**
30:     **end for**
31:   **end for**
32: **end for**
33: **return** BestSentence

We have a three-dimensional search, starting from creation of *nPop* particles that will have the number *nVar* of the suggested candidate solutions and will, in turn, have the generated words for the translation. This is found the crucial task performed by the most internal cycle in which, the proposed candidate words are generated and evaluated to find the best candidate equivalent for the word translation. If the generated word is exactly the same as the target word, this word is added to the best sentence until the best sentence is completed, or the final condition of the cycles is met. At the end of the PSO algorithm cycle, the correct as most similar to the reference sentence will be produced. In order to evaluate the similarity at the word or the sentence level, we use the cosine similarity function that computes the similarity range of the values, and depending on the result, we constantly update the velocity values to accelerate the search process.

3   Results and Discussion

In our PSO algorithm, many particles are working at the same time to find the best solution. The algorithm permits to have n solutions inside every particle, which transforms into the conclusion that increasing the probability of finding the solution is achieved in a shorter period of time and using a fewer number of the resources. Communication between the particles permits them to communicate the other particles when the best solution has been found. The same solution applies to a multilingual translator since our solution can have the same sentences translated in every incoming source language. This way streaming translation can be processed. As shown in Tables 2 and 3, the encoded word is independent of the language meaning our technology is of a universal nature.

Reducing complexity is more important for complete sentences rather than for words only because in general, each language has well-defined sentence structures such as "good morning", whose translations into other languages can be performed without any predefined linguistic analysis. For this reason, we suggest using sentence identifiers that can be paired up. Below, some examples are presented in three different languages. As shown in Table 4, it is the same identifier applied to three different languages.

Table 4. Sentence identifiers in three languages.

Language	Sentence identifier	Sentence
English	5	It is informal
Polish	5	To jest nieformalne
Spanish	5	Es informal

It is possible to gain an easy access to the phrases or language rules such as acronyms, which are sometimes the same in all languages, regardless of the ways in which they are written. Below is an example of an acronym UK in three languages (Table 5).

Table 5. Acronym identifiers in three languages.

Language	Acronym identifier	Acronym	Translation
English	6	UK	United Kingdom
Polish	6	UK	Wielka Brytania
Spanish	6	UK	Reino Unido

Some of the cost functions do not need very complex solutions. For example, in the model below we associate the cost of the code with the *WordId* and an absolute value of subtraction of *GeneratedWordID* from *TargetWordID* turned out to be sufficient as the only one.

$$CostWordIDCode = |\ TargetWordIdCode - GeneratedWordId - Code|\quad (1)$$

Cosine similarity compares similarity between two encoded sentences and between two encoded words to update the velocity rules and improve the search task.

With the granularity increase, the PSO becomes more and more complex both to design it and to apply for any purpose.

As many authors conclude, translation at the sentence level seems to be an easier task, regardless of the syntax and grammar rules. In this work, we ignore those rules for the moment.

Adapting the PSO algorithm improves understanding of a particular language rules such as syntax and grammar.

4   Conclusions

This paper presents a new approach to the machine translation. It applies a PSO algorithm in order to find the best translation of a set of some complex sentences. The best translation is defined as the one that is the most similar to the reference sentence. Our objective function was cosine similarity as a cost function to determine mapping of words and sentences which are the most similar to each other.

Application of the PSO algorithm improved our translation precision for encoded sentences and encoded words owning to using the same identifier for all languages.

In the future, we plan to apply language rules such as syntax and grammar. Addressing language granularity causes increase in translation complexity exponentially because due to the higher probability of word combinations in the source and target languages. It is thus necessary to explore a wide range of the same sentences or the same words for their different meanings.

Acknowledgements. The project is supported by a research grant No. DSA/103.5/16/10473 awarded by PRODEP and by Autonomous University of Ciudad Juarez in Mexico. Title - Detection of Cardiac Arrhythmia Patterns through Adaptive Analysis.

References

1. Hutchins, W.J.: Machine translation: a brief history. In: Concise History of the Language Sciences: From the Sumerians to the Cognitivists, pp. 431–445 (1995)
2. Eiben, A.E., Smith, J.E.: Introduction to Evolutionary Computing, 2nd edn. Springer, Heidelberg (2015)
3. Otto, E., Riff, M.C.: EDA: an evolutionary decoding algorithm for statistical machine translation. Appl. Artif. Intell. **21**(7), 605–621 (2007)
4. Ameur, D., David, L., Kamel, S.: Genetic-Based Decoder. Lecture Notes in Computer Science (2016)
5. Menai, M.E.B.: Word sense disambiguation using evolutionary algorithms – application to Arabic language. Comput. Hum. Behav. **41**, 92–103 (2014)

6. Mihalcea, R., Corley, C., Strapparava, C.: Corpus-based and knowledge-based measures of text semantic similarity. In: American Association for Artificial Intelligence, pp. 775–780 (2006)
7. Dehak, N., Dehak, R., Glass, J., Reynolds, D., Kenny, P.: Cosine similarity scoring without score normalization techniques. In: De Odyssey 2010, Brno (2010)
8. Kazemi, A., Toral, A., Way, A., Monadjemi, A., Nematbakhsh, M.: Syntax- and semantic-based reordering in hierarchical phrase-based statistical machine translation. Expert Syst. Appl. **84**, 186–199 (2017)
9. Choi, H., Cho, K., Bengio, Y.: Context-dependent word representation for neural machine translation. Comput. Speech Lang. **45**, 149–160 (2017)

Experimental Investigation of Ant Supervised by Simplified PSO with Local Search Mechanism (SAS-PSO-2Opt)

Ikram Twir[1], Nizar Rokbani[1,2,4(✉)], Abdelkrim Haqiq[3],
and Ajith Abraham[4]

[1] ISSAT Sousse, University of Sousse, Sousse, Tunisia
{touir.ikram.1993,nizar.rokbani}@ieee.org
[2] REGIM-Lab, National Engineering School of Sfax,
University of Sfax, Sfax, Tunisia
[3] FST, Hassan 1st University, Settat, Morocco
abdelkrim.haqiq@uhp.ac.ma
[4] Machine Intelligence Research Labs (MIR Labs),
Scientific Network for Innovation and Research Excellence,
P.O. Box 2259, Auburn, Washington 98071, USA
ajith.abraham@ieee.org

Abstract. Self-adapting heuristics is a very challenging research issue allowing setting a class of solvers able to overcome complex optimization problems without being tuned. Ant supervised by PSO, AS-PSO, as well as its simplified version SASPSO was proposed in this scope. The main contribution of this paper consists in coupling the simplified AS-PSO with a local search mechanism and its investigations over standard test benches, of TSP instances. Results showed that the proposed method achieved fair results in all tests: find the best-known solution or even find a better one essentially for the following cases: eil51, berlin52, st70, KroA100 and KroA200. The proposed method turns better results with a faster convergence time than the classical Ant Supervised by PSO and the standard Ant Supervised by PSO as well as related solvers essentially for eil51, berlin52, st70 and kroA100 TSP test benches.

Keywords: PSO · ACO · TSP · Ant Supervised by PSO

1 Introduction

The Traveling Salesman Problem is NP-Hard problem. It consists of visiting all cities once using the shortest path. TSP is a common test bench for computational intelligent techniques, essentially those focusing on combinatorial NP-hard problems. The TSP was largely investigated using Ant colony optimization and related techniques. The ANT paradigm is inspired from the natural ant swarm behavior and collaboration capacity allowing to the group to achieve complex collective goals while involving very limited individual resources [1, 2]. Classically in ANT optimization schema, the heuristic runs several times and the best obtained result is assumed to be the solution, it is also possible to implement a self-adapting mechanism allowing to ANT to self-tune

© Springer International Publishing AG, part of Springer Nature 2018
A. Abraham et al. (Eds.): SoCPaR 2017, AISC 737, pp. 171–182, 2018.
https://doi.org/10.1007/978-3-319-76357-6_17

its parameters [3, 4] for single and multi-objective optimization. Tuning Ant, or any heuristic solver, parameters using another heuristic technique such as the particle swarm optimization [5], PSO, is hybridization schema allowing to generate self-adaptive meta-heuristics; such a technique was proposed in [6–8] to solve the TSP and in [9] for multiprocessor job scheduling. Ant Supervised by PSO is a meta-heuristic combining ACO and PSO where PSO is used to tune the ACO parameters [6–8], All these methods are based on the same hybridization schema. Authors investigations covered the impact of developed PSO and ACO heuristics using essentially the TSP test benches. In [5], (Elloumi et al. 2009) proposed the ant supervised by PSO (AS-PSO) for the first time. They explained the main goal of the hybridization and algorithm basics where the PSO is used to find ACO parameters. (Rokbani et al. 2013) proposed AS-PSO to solve a traveling salesman problem [7], however, authors did not use the standard Traveling salesman problem; they use Tunisian cities to test the algorithm performances. In [8], Kefi et al. investigate the inertia weight PSO to tune ACO parameters, the ant supervised by PSO for the standard TSP test benches with the use of a local search policy (2Opt). In [6], Rokbani (2013) proposed new version of AS-PSO, which is the simplified AS-PSO. The proposal is based on ACO and the simplified PSO (2010) [9]. The proposal is used to solve Tunisian cities, not the standard TSP test benches without using a local search policy (2opt). ACOMAC is also a hybrid method based on ant colony optimization with multiple ant clans dedicated to large TSP problems, which is based on multiple ant clans' and genetic algorithm. The ACOMAC algorithm is then combined with multiple nearest-neighbors (NN) and the dual nearest neighbor (DNN) to enhance its performance [10]. A Neuro-Immune Network, RABNET-TSP is a self-organized neural network with adapted immune systems named "real valued antibody network" for solving the Traveling Salesman Problem [11]. In [12], Massuti contribute with two modifications in the original RABNET-TSP the first one is about adding of a threshold to activate the kernel function and the second one is the winners' stabilization mechanism. For larger TSPs instances, this modified algorithm converges more quickly than the previous version but not with best solutions. An improved ACO for generalized TSP is presented in [13], where the generalized TSP, GTSP, is composed of a set of TSP instances using a cluster modeling approach in which tours are grouped in clusters around city prior to linking the cities clusters between each other. A heuristic method consisting in removing a "cross line" used by an ant to reduce the local search in ACO [14]. A hybridization of Ant Colony Optimization and Genetic algorithm is proposed in [15], the cooperative genetic ant system, CGAS, where a cooperation between GA and ACO allowing to suggest to ACO a better proposition for next iteration and helping ACO in getting trapped in local optimum [15]. In [16], authors combine two meta-heuristics, which are Water flow algorithm, WFA, and Tabu Search, TS, to provide better solutions for symmetric TSP problems. The main goal of the WFA is the exploration and TS is the generation of best solution where it has a memory to register the previous solution and use it to find the best one in the next attempts. In [17], Mahia et al. (2015) proposed a close variant solving the traveling salesman problem, which is the ant supervised by PSO with a local search 3Opt. In this paper, we will implement Simplified AS-PSO to solve the standard traveling salesman problem test benches with the use of Local search, 2Opt, that is not developed previously. Our contribution is based on a self-adaptive ant based on a specific PSO variant, the simplified PSO, SAS-PSO is coupled with a local heuristic search allowing of avoiding limited local optimums. The proposal is applied to the main TSP test benches and compared to main contributions. The remaining of this paper is organized as follows: Sect. 2 presents the methods and techniques used in this paper, Sect. 3 presents the experimental methodology as well as the used test benches and the obtained results followed by the comparison with related work. The paper ends by a conclusion and perspective works, Sect. 4.

2   Methods and Techniques

2.1   Ant Colony Optimization ACO

ACO Stands for Ant Colony Optimization and was proposed by Dorigo et al. in [1] as a direct inspiration form natural ants organization. It is based on ant's behavior. This method is based on ant collaboration strategy based on pheromone to find the shortest tour to their food source. Communication strategy consists in the use of biological marker to characterize the path with probability leads to the next step. To calculate the probability $P_{i,j}^k$ ants use Eq. (1) where $\tau_{i,j}$ represents the substrate quantity between node j and i, Ω_i denotes ith neighborhood, α, β are pheromone parameters. $P_{i,j}^k$ Is standing for the probability of k_{th} ant passing the arc (i, j), see Fig. 1.

$$P_{i,j}^k = (\tau_{i,j}^{k-1})^\alpha * \eta_{i,j}^\beta / \sum_{j \in \Omega i} (\tau_{i,j}^{k-1})^\alpha * \eta_{i,j}^\beta \qquad (1)$$

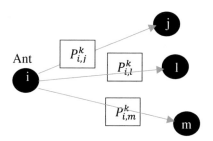

Fig. 1. Ant search strategy

Ants naturally move following the path with the highest probability P_{ij}^k, P_{il}^k, and P_{im}^k, when moving from a city i to city j.

To search for their food, natural ants start by their home city to attend the food place and return to the first point with the shortest path assumed the one where the maximum quantity of pheromone is available. To update the pheromone ants use Eq. (2) Where ρ is the Pheromone decay coefficient [18].

$$if\,(i,j)\epsilon\,Best\,\,Tour\,\,\tau_{ij} = (1-\rho)\tau_{i,j}^{k-1} + \rho\Delta_{ij}^{k}$$
$$else\,\,\tau_{i,j}^{k-1} = \tau_{i,j}^{k-1} \tag{2}$$

2.2   PSO and Simplified PSO

Particle Swarm Optimization, PSO, is a very popular and effective computational intelligence algorithm proposed by (Eberhart and Kennedy 1995) [19, 20]. It is inspired from the collective and the social intelligence of bird flocks, fish banks and similar biological swarms, where each individual, called particle is informed about the solution found by the community as well as the solution in his close neighborhood to update its own position as in Eqs. (3) and (4). Where w stands for inertia weight, C1, and C2 are respectively the cognitive and the social coefficients, P_{lbest}, P_{gbest} are the best local and global positions.

$$v_i = wv_i + C1 * rand() * (P_{lbest} + x_i)$$
$$+ C2 * rand() * (P_{gbest} + x_i) \tag{3}$$

$$x_i = x_i + v_i \tag{4}$$

The simplified PSO is a swarm focusing only on the social factor and ignoring completely the cognitive factor represented by the neighborhood [9], the PSO-VG particles displacements are ruled by Eqs. (5 and 6).

$$v_i = wv_i + C2 * rand() * (P_{gbest} + x_i) \tag{5}$$

$$x_i = x_i + v_i \tag{6}$$

This simplified AS-PSO gives closet solutions with an optimum path equal to the best global Tour, see Eq. 7. Similar proposal to PSO-VG is the PSO-G where the variant updates its position with only global learning solution, see Eqs. 5 and 6.

$$X = \sum_{i=1,j=2}^{N} d_{ij} \tag{7}$$

Where d is the distance between city i and city j and N is the total cities number.

2.3   Local Search 2Opt

2Opt is a well-known local search policy. It consists in removing two cross lines from a graph between three cities. 2Opt is a specific variant from K-Opt, which is proposed in [22] by (Croes 1985). K-Opt is a fast local search algorithm applied to solve optimization problem; it consists in removing k arcs from the current node and replace them in other positions. The two well-known KOpt particular cases are 2Opt and 3Opt. K-Opt is an iterative algorithm incorporated into a heuristic algorithm to help him in the search process without missing its structure. 2-opt algorithm [23] consists in removing two connections from the current node and reconnect them to form another path without missing the tour construction. It is valid only if the new tour is shorter than the recent one. There is only one way to reconnect edges. This loop continue the execution until any improvement will be done.

2.4   Simplified Ant Supervised by PSO-Local Search

The simplified ant supervised by PSO, SAS-PSO version was proposed by (Rokbani et al. 2013) [6]. SAS-SPO in general, SPSO is running to control heuristic ACO Parameters while ACO is used to solve the main problem, in our case TSP [7]. Classically, the user have to fix some parameters by generating many tests to choose the performing ones. SAS-PSO avoids the local solution and run directly to find the global one, hence, the simplified AS-PSO is a global search algorithm. It uses only social behavior to search for the best solution. PSO will use the pheromone to enhance the global search process. Its particles form the algorithm solutions meanwhile each PSO particle corresponds to a triple (α, β, ρ). PSO particles move in the search space using their velocities to update their positions basing on Eqs. (5) and (6). Velocity depends only from the global solution, means the cognitive coefficient is removed. The proposal is based on simplified particle swarm optimization, ant colony optimization with local search 2Opt applied for the standard TSP, which is a novel contribution and was not developed previously. To avoid being trapped in local optimum the SAS-PSO uses a local search heuristic policy consisting in removing cross lines over three cities and replacing them by the shortest path between the three cities based on classical tour length strategy. The proposed method architecture is illustrated with Fig. 2. The proposed method flowchart is presented with Fig. 3. Algorithm 1 presents the proposed method.

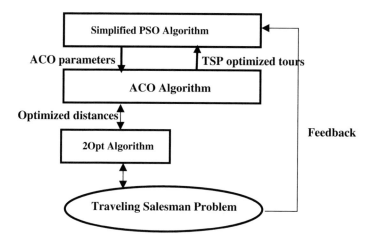

Fig. 2. Simplified AS-PSO-2Opt architecture

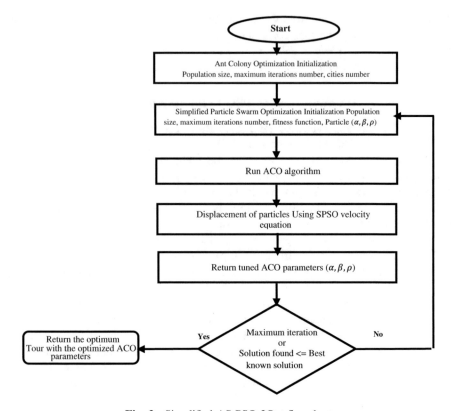

Fig. 3. Simplified AS-PSO-2Opt flowchart

First, we have to initialize the ACO parameters (swarm size, maximum iteration, cities number) and the SPSO Settings (Swarm size, C2, W, fitness function, Particle (α, β, ρ)). Then, ACO algorithm is running to solve the TSP problem with initial parameters generated by initial SPSO algorithm. The next step is updating ACO parameters using SPSO velocity and position equations by SPSO algorithm, means displacement of SPSO particles. Hence, ACO is running to evaluate new parameters by trying to solve the main problem. SPSO is running until a maximum iteration or a global solution equal or better than the best-known solution is reached.

3   Experimental Results

3.1   Experimental Protocol

Experimental results are obtained using Matlab software, R2013a. It runs on desktop machine with ACER Intel Core i7, 8 GB RAM size. To study the algorithm performances basing on a statistical analysis, we have to find the average solution, the standard deviation and variance of each solution for each test bench relative to the best-known solution. To qualify obtained solutions we have to calculate the error for each result basing on Eq. (8), which is used in [7, 8]. BKS stands for the best-known solution and Avg is the average solution.

$$Error(err.) = ((Avg - BKS) \div BKS) * 100 \qquad (8)$$

Several test benches are used from the standard library, TSPLib. [21] such as eil51, berlin52, st70, st76, rat99, kroA100, eil101, ch150 and kroA200. Each test bench is characterized by its best Known solution. Many attempts are done to evaluate this algorithm (SAS-PSO-2Opt), Statistical analysis are applied for TSP for 10000 executions.

Our contribution consists in testing the SAS-PSO-2Opt by conducting two experimentations. Our objective function is minimizing the total distance d between N cities as Eq. 9:

$$\min(\sum_{i=1,j=2}^{N} d_{i,j}) \qquad (9)$$

The first one set ant population number to 20. The second experimentation use an ant population equal to city number. For the SAS-PSO, the velocity depends only from the global coefficient C2 meanwhile we will test the variation of C2 and its influence in the algorithm. In our investigation, PSO social coefficient is varied between 0.2 and 2.8 with a step equal to 0.4. The maximum iterations used by PSO is equal to 100. PSO was executed with particles population equal to 10.

3.2   Experimental Results

Figures 4, 5, 6, 7, 8, 9, 10 and 11 present the shortest path obtained by eil51, berlin52, st70, eil76, kroA100, eil101, ch150 and finally kroA200. The best Tour obtained by eil51 simulation is 442, which is illustrated in Fig. 4. The shortest path attended by berlin52 execution is 7935, which is presented with Fig. 5. Figure 6 shows the best result obtained my st70 test bench that is equal to 754. The optimum path obtained by eil76 is 579; it is illustrated in Fig. 7. The best solution obtained by eil101 is 695, which is presented with Fig. 8. Figure 9 presents the best tour length of kroA100, which is 23181. 7051 is the optimum tour obtained by ch150 test bench and it is showed in Fig. 10. Figure 11 shows the optimum solution of kroA200 that is equal to 31276. Table 1 illustrates the shortest path ever found with the best parameters. Eil51 shortest path is 426 with PSO social coefficient equal to 2.4 and ACO settings correspondent to $\alpha = 0.97754$, $\beta = 5$, $\rho = 0.21663$. The optimum solution given for SAS-PSO by berlin52 is 7542 with C2 = 2.4 and $\alpha = 0.5$, $\beta = 3.2138$, $\rho = 0.27447$. The best global solution for st70 is 675 with C2 = 0.8 and $\alpha = 1.9836$, $\beta = 3.479$, $\rho = 0.026371$. Eil76 shortest tour is 543 with C2 = 2 and $\alpha = 1.5119$, $\beta = 4.2564$, $\rho = 0.41424$. The best optimum for eil101 is 645 with C2 = 2.8 and $\alpha = 1.7963$, $\beta = 4.7321$, $\rho = 0.22427$. The average of kroA100 is 21305 with C2 = 0.8 and $\alpha = 1.6549$, $\beta = 3.3073$, $\rho = 0.32603$. The test bench ch150 best solution is 6606 with C2 = 1.2 and $\alpha = 0.95006$, $\beta = 3.8297$, $\rho = 0.21841$. KroA200 best solution is 29997 with bellowed parameters; C2 = 2 and $\alpha = 0.5$, $\beta = 4.6791$, $\rho = 0.041289$. Table 1

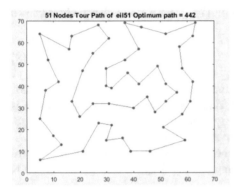

Fig. 4. eil51 shortest path using SAS-PSO

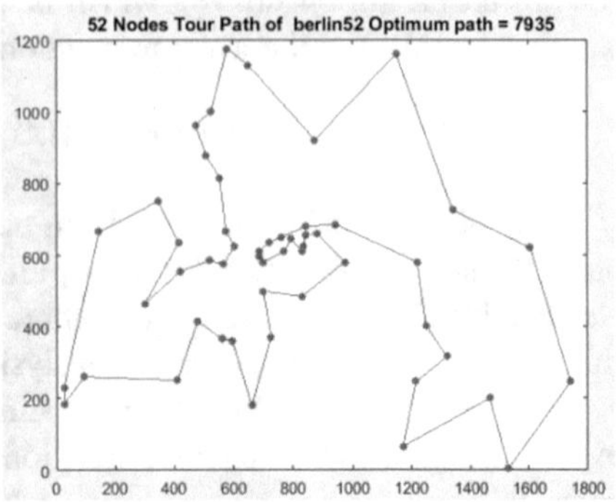

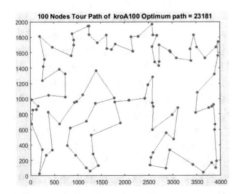

Fig. 5. berlin52 shortest path using SAS-PSO

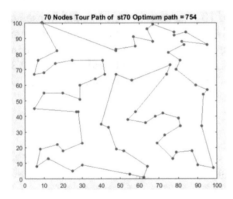

Fig. 6. st70 shortest path using SAS-PSO

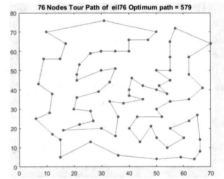

Fig. 7. eil76 shortest path using SAS-PSO

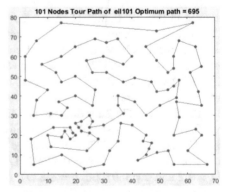

Fig. 8. KroA100 shortest path using SAS-PSO

Fig. 9. eil101 shortest path using SAS-PSO

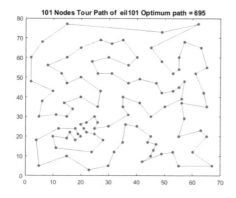

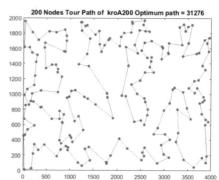

Fig. 10. eil101 shortest path using SAS-PSO

Fig. 11. kroA200 shortest path using SAS-PSO

Table 1. Optimized ACO parameters with ant num = city number

TSP test benches (Problems)	Statistics	Best results found by SAS-PSO	Optimized parameters	BKS
eil51	Avg.	426	$\alpha = 0.97754$	426 [17]
	SD	9.6403	$\beta = 5$	
	Error (%)	0	$\rho = 0.21663$	
	ANT Num	20		
	C2	2.4		
	Time (s)	1883.653		
berlin52	Avg.	7542	$\alpha = 0.5$	7542 [8]
	SD	206.1429	$\beta = 3.2138$	
	Error (%)	0	$\rho = 0.27447$	
	ANT Num	52		
	C2	2.4		
	Time (s)	4339.313		
st70	Avg.	675	$\alpha = 1.9836$	675 (*)
	SD	20.6865	$\beta = 3.479$	
	Error (%)	0	$\rho = 0.026371$	
	ANT Num	20		
	C2	0.8		
	Time (s)	3374.2567		
eil76	Avg.	543	$\alpha = 1.5119$	538 [17]
	SD	13.1426	$\beta = 4.2564$	
	Error (%)	0.92937	$\rho = 0.41424$	
	ANT Num	20		
	C2	2		
	Time (s)	5969.1875		

(continued)

Table 1. (*continued*)

TSP test benches (Problems)	Statistics	Best results found by SAS-PSO	Optimized parameters	BKS
kroA100	Avg.	21305	$\alpha = 1.6549$	21282 (*)
	SD	674,4597	$\beta = 3.3073$	
	Error (%)	0,10807	$\rho = 0.32603$	
	ANT Num	20		
	C2	0.8		
	Time (s)	7724,335488		
Eil101	Avg.	645	$\alpha = 1.7963$	629 (*)
	SD	12.1358	$\beta = 4.7321$	
	Error (%)	2.5437	$\rho = 0.22427$	
	ANT Num	101		
	C2	2.8		
	Time (s)	9998.2011		
ch150	Avg.	21305	$\alpha = 1.6549$	6528 (*)
	SD	674,4597	$\beta = 3.3073$	
	Error (%)	0,10807	$\rho = 0.32603$	
	ANT Num	20		
	C2	0.8		
	Time (s)	17052.594117		
kroA200	Avg.	6606	$\alpha = 0.95006$	29368 (*)
	SD	176.5837	$\beta = 3.8297$	
	Error (%)	1.1949	$\rho = 0.21841$	
	ANT Num	20		
	C2	1.2		
	Time	32715.957340		

Table 2. SAS-PSO-2opt convergence time

	Eil51	Berlin52	St70	Eil76	kroA100	Eil101	Ch150	kroA200
Time fitting (s)	1883.653	4339.313	3374.2567	5969.1875	7724,335488	9998.2011	17052.594117	32715.957340
Time (s)	1.883653	4.339313	3.3742567	5969.1875	7.724335488	9.9982011	17.052594117	32.715957340

column 4, illustrated the best parameters tuned by SPSO for each test bench. (*) Test bench where SAS-PSO-2OPT find better than existing solutions.

ACO self-tuned parameters are selected at intervals, intervals belong to:

- $\alpha \in [0.5\,2], \beta \in [3\,5], \rho \in [0\,0.35]$

4   Conclusion

In this paper a new self-adaptive Ant supervised by PSO solver is proposed, compared to the native version of SAS-PSO the new proposal implements an additional local search policy based on 2-Opt. The proposal was validated using the TSPlib test benches and returned better results than the best known solutions, BKS, in the following cases: eil51, berlin52, st70, eil76, kroA100, ch150, and kroA200, see Table 2 for the time speed convergence and Table 1 for comparative results. SAS-PSO-2Opt showed also fair results in large TSP problems such as KroA 100 & 200. The proposed method is given better results that the classical AS-PSO and the standard AS-PSO. Further investigations will focus on Fuzzy-AS-PSO, proposed in [7].

References

1. Dorigo, M., Birattari, M., et al.: Swarm intelligence. Scholarpedia **2**(9), 1462 (2007)
2. Dorigo, M., Di Caro, G.: Ant colony optimization: a new meta-heuristic. In: Proceedings of the 1999 Congress on Evolutionary Computation-CEC 1999, vol. 2 (1999)
3. Förster, M., Bicke, B.: Self-Adaptive Ant Colony Optimisation Applied to Function Allocation in Vehicle Networks (2007)
4. Ying, W., Jianying, X.: An adaptive ant colony optimization algorithm and simulation. Acta Simulata Systematica Sinica **1**, 009 (2002)
5. Elloumi, W., Rokbani, N., Alimi, A.M.: Ant supervised by PSO. In: Proceedings of International Symposium on Computational Intelligence and Intelligent Informatics, pp. 161–166 (2009)
6. Rokbani, N., Abraham, A., Alimi, A.M.: Fuzzy ant supervised by PSO and simplified ant supervised PSO applied to TSP. In: The 13th International Conference on Hybrid Intelligent Systems, HIS 2013, Gammarth, Tunisia, 4–6 December 2013. IEEE (2013). ISBN 978-1-4799-2438-7
7. Rokbani, N., Momasso, A.L., Alimi, A.M.: AS-PSO, ant supervised by PSO meta-heuristic with application to TSP. In: Proceedings Engineering and Technology, vol. 4, pp. 148–152 (2013)
8. Kefi, S., Rokbani, N., Krömer, P., Alimi, A.M.: Ant supervised by PSO and 2-OPT algorithm, AS-PSO-2Opt, applied to traveling salesman problem. In: IEEE International Conference on System Man and Cybernetics SMC (2016)
9. Pedersen, M.E.H., Chipperfield, A.J.: Simplifying particle swarm optimization. Appl. Soft Comput. **10**, 618–628 (2010)
10. Cheng-Fa, T., Chun-Wei, T., Ching-Chang, T.: A new hybrid heuristic approach for solving large traveling salesman problem. Inf. Sci. **166**(1), 67–81 (2004)
11. Pasti, R., de Castro, L.N.: A neuro-immune network for solving the traveling salesman problem. In: International Joint Conference on Neural Networks, IJCNN 2006. IEEE (2006)
12. Masutti, T.A.S., de Castro, L.N.: A self-organizing neural network using ideas from the immune system to solve the traveling salesman problem. Inf. Sci. **179**, 1454–1468 (2009)
13. Jun-man, K., Yi, Z.: Application of an improved ant colony optimization on generalized traveling salesman problem. Energy Procedia **17**, 319–325 (2012)
14. Junqiang, W., Aijia, O.: A hybrid algorithm of ACO and delete-cross method for TSP. In: The IEEE International Conference on Industrial Control and Electronics Engineering, pp. 1694–1696 (2012)
15. Dong, G.F., Guo, W.W., Tickle, K.: Solving the traveling salesman problem using cooperative genetic ant systems. Expert Syst. Appl. **39**, 5006–5011 (2012)
16. Othman, Z.A., Srour, A.I., Hamdan, A.R., Ling, P.Y.: Performance water flow-like algorithm for TSP by improving its local search. Int. J. Adv. Comput. Technol. **5**, 126–137 (2013)

17. Mahia, M., Baykanb, Ö.K., Kodazb, H.: A new hybrid method based on particle swarm optimization, ant colony optimization and 3-OPT algorithms for traveling salesman problem. Appl. Soft Comput. **30**, 484–490 (2015)
18. Dorigo, M., Gambardella, L.M.: Ant colony system: a cooperative learning approach to the traveling salesman problems. Technical report TR/IRIDIA/1996-5, IRIDIA, Université Libre de Bruxelles (1997)
19. Kennedy, J., Eberhart, R.: Particle swarm optimization. In: IEEE International Conference on Neural Networks, pp. 1942–1948 (1995)
20. Rokbani, N., Alimi, A.M.: Inverse kinematics using particle swarm optimization, a statistical analysis. Procedia Eng. **64**(Suppl. C), 1602–1611 (2013). https://doi.org/10.1016/j.proeng.2013.09.242
21. Reinelt, G.: TSPLIB—a traveling salesman problem library. ORSA J. Comput. **3**(4), 376–384 (1991). https://doi.org/10.1287/ijoc.3.4.376
22. Croes, G.A.: A method for solving traveling salesman problems. Oper. Res. **6**, 791–812 (1958)
23. Dorigo, M., Stutzle, T.: Ant Colony Optimization, Massachusetts Institute of Technology (2004)

A Human Identification Technique Through Dorsal Hand Vein Texture Analysis Based on NSCT Decomposition

Amira Oueslati[1(✉)], Nadia Feddaoui[2], and Kamel Hamrouni[1]

[1] LR-SITI Laboratory, National Engineering School of Tunis,
University ELManar, Tunis, Tunisia
amiraoueslati@yahoo.fr, Kamel.hamrouni@enit.rnu.tn
[2] LR-SITI Laboratory, ISD, University Manouba Tunis, Manouba, Tunisia
nadia.feddaoui@insat.rnu.tn

Abstract. Dorsal hand vein identification has been recently given greater attention in human recognition and it's becoming increasingly an active topic in research. This paper presents a personal identification method based on dorsal hand vein texture. The Method includes four steps, in the first one, pre-processing phase is applied on the image contrast in order to produce a better quality of dorsal hand vein image, then region of interest (ROI) is extracted, in the second step, we have proposed a novel encoding method based on Nonsubsampled contourlet transform (NSCT) and phase response information then we divided the resulting image into local region, and statistical descriptors are calculated in each block in order to reduce the size of the characteristic vector and create a code of 512 bytes. Then, we computed the modified Hamming distance between templates to find out the similarity between two dorsal hand veins.

The method is tested on the "GPDSvenasCCD" database. The experimental results illustrate the effectiveness of this coding in Identification mode of biometric dorsal hand vein: 99.96% of rank-one recognition rate. Therefore, the coding process is presented to achieve more satisfactory results than performed by traditional statistical based approaches. The performed numerical results prove the robustness of our approach to extract discriminative features of dorsal hand veins texture, which suggests a significant advance in texture Identification.

Keywords: Dorsal hand veins identification
NonSubsampled Contourlet Transform (NSCT) · ROI extraction
Feature extraction · Statistical descriptors · Multi-block · Phase response

1 Introduction

Biometrics is used in the computer sciences to refer the domain of mathematical analysis regarding unique human features. Dorsal hand vein biometrics is a new type of biometric system. In fact, this modality presents a richer distinctive pattern.

Biometric identification systems based on dorsal hand vein patterns are becoming increasingly important as they contain properties like universality, stability, uniqueness

© Springer International Publishing AG, part of Springer Nature 2018
A. Abraham et al. (Eds.): SoCPaR 2017, AISC 737, pp. 183–193, 2018.
https://doi.org/10.1007/978-3-319-76357-6_18

and strong immunity to forgery. Since the veins lie underneath the skin and are, in most cases, not visible to the naked eye, they provide a strong resistance against forgery. The complex vascular pattern present inside the hand allows the computation of a good set of features that can be used for personal identification Anatomically, the shape of vein in the dorsal of the hand is distinct from each other [1–3]. These feature makes it a more reliable biometric for personal identification [4]. Furthermore, the state of skin, temperature and humidity has little effect on the vein image, unlike other module like fingerprint and face. The hand vein biometrics principle is non- invasive in nature where dorsal hand vein pattern are used to verify the identity of individuals [5]. In this paper, to obtain a high-performance of dorsal hand veins recognition system, we have applied the Nonsubsampled Contourlet Transform (NSCT) method which is a shift-invariant, multi-scale, and multi-directional transform. It can capture significant veins features along all directions.

This paper is structured as follows. In Sect. 2, existing methods in literature are briefly reviewed. In Sect. 3, the proposed dorsal hand veins identification method using the NSCT is presented. In Sect. 4, experimental results are given and discussed. Finally, conclusions are drawn.

2   Review of Some State of Art

Human identification has been evaluated through dorsal hand veins images with high recognition rate approaches. Among biometric features, dorsal hand veins are often applied in both field researches and industry. For these reasons, several techniques are presented. Image is a set of details which appear at different resolutions. For that, multi-resolution decomposition is used to analyse different type of image's structure, among them there is the Hough transform, the Gabor filter, discrete Curvelet transform, Ridgelet transform and discrete wavelet transforms. In [6] they used the Median filter as preprocessing step and then they applied Gabor filter as feature extraction method they found 1.41% as EER result.

The work [7] which is based on discrete wavelet transform, authors extract region of interest, applied a double adaptive equalization contrast to accentuate the vein contrast then applied their new adaptive feature extraction method for the dorsal hand vein biometrics, they tested their method on SAB11 Database. In [8] authors proposed to use an infrared camera to capture the dorsal hand image. Then it applied a method known as maximum curvature to find the center lines of veins and finally for matching and decision, it used a simple correlation method. The authors in [9] have applied a new vein recognition approach based on adaptive Hidden Markov Model (HMM), they used stepper increasing approach according to different databases to optimize the parameters of HMM, and in each database every vein object can be represented as a HMM. We find also the approach based on curvelet [10], authors used here low-cost camera for acquisition images, then region of interest was defined experimentally by media, they extract the largest possible area of the dorsal side of the hand, the curvelet transform is applied for decomposition as characterisation and finally they used the Random Forest classifier, implemented in the WEKA platform.

We can find other methods based on triangulation of minutiae, Knuckle tip…etc. In [11] authors extract the minutiae features from the dorsal venous patterns, they included the end points and the distance between two end points as measured along the boundary of the image. The end-points-tree (EP-tree) is proposed to accelerate the matching performance and evaluate the discriminating power of these end points for person recognition method. In [12] we identify the approach based on Firefly Algorithm and Knuckle tip (control points), the authors used the vein intersections and knuckle shape features.

In this paper we have also presented identification performance for dorsal hand vein identification from multi-resolution approach based on Non-Subsampled Contourlet Transform and Tests are carried out on GPDSvenasCCD database [13].

3   Proposed Method

Biometric systems are composed on four steps which are:

- Pre-processing: The goal of this step is to improve the quality image and suppresses unwilling distortions, enhance image features.
- ROI extraction: The aim of this step is to locate the region of interest of the dorsal hand image which contains blood vein patterns; it allows extracting the most informative area of dorsal hand image. So, it reduces a lot of useless data without losing much useful information.
- Texture characterization: It aims to extract most informative descriptors from texture and create a vector feature.
- Comparison and decision: This step consists on matching the obtained dorsal hand vein feature vectors to take the best decision of recognition.

Figure 1 shows the process of the identification model using dorsal hand veins biometrics that we use in our approach of identification.

Fig. 1. The process of the dorsal hand veins recognition.

As pre-processing stage, an anisotropic diffusion filter is applied on the dorsal venous images to preserve edges, for ROI extraction we detect Boundary and find the contour of the dorsal hand region by the inner border tracing algorithm [14], we locate center of gravity of image, and finally a square region of size R × R pixels is extracted. We have applied NSCT decomposition for image, then phase information response is applied to determine the dominant orientation.

ROI resulting after decomposition is divided into multi-block and statistical descriptors are calculated to reduce feature vector size.

Finally hamming distance is calculated to find similarity into images in order to take a decision.

3.1 Dorsal Hand Veins Feature Extraction

In a previous work [15] we have extract the region of interest of dorsal hand vein image and we have been described the process step by step, the result of our ROI extraction method is shown below in Fig. 2.

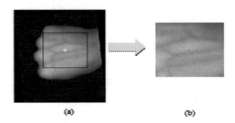

(a)   (b)

Fig. 2. ROI extraction: (a) Square region detection (b) Final ROI.

Now, region of interest is extracted so we must analyse the dorsal hand vein texture to detect the most distinctive features that help us to identify the person. In our method, we will apply the phase response of NonSubsampled Contourlet Transform to extract texture features.

3.2 Texture Description Based on NSCT

Dorsal Hand Vein Texture Description
As shown in Fig. 2, the red blood cells in the dorsal hand veins show up on the map of the hand as black lines, whereas the hand structure shows up as white. This vein pattern is then verified to authenticate the individual.

There are mainly two types of dorsal hand veins, namely cephalic and basilic. The basilic veins are the group of veins attached with surface of hand. This part of hand is rich of blood vessels which are well structured, homogeny, and directional, represented a fine detail, linear and interconnected. Dorsal veins are considered as a macrotexture samples being automatically extracted from the user's hand image.

NSCT Description
In 2005, Do et al. [16] proposed the NSCT which is a shift-invariant version of the contourlet, and multidirectional expansion that has a fast implementation. The NSCT eliminates the down-samplers and the up-samplers during the decomposition of the image.

Instead, it is built ahead the nonsubsampled pyramids filter banks (NSPFBs) and the nonsubsampled directional filter banks (NSDFBs), the first one ensures the multi-scale property [16], and the second one provides the directionality the NSPFB, employed by the NSCT, is a two-channel nonsubsampled filter bank (NFB).

The NSCT is obtained by carefully combining the NSPFB and the NSDFB [16], as shown in Fig. 3.

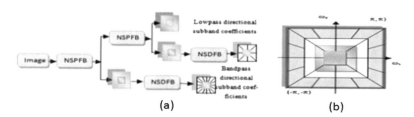

(a)   (b)

Fig. 3. Nonsubsampled contourlet transform structure: (a) NSFB that implements the NSCT [21], (b) Resulting 2-D frequency division [21]

Figure 3(a) shows the structure of the two-channel NSFB, and Fig. 3(b) shows a 2-D frequency domain divided into a number of subbands. The NSPFB consists on a high-pass (HP) subband and a low-pass (LP) subband, and the NSDFB decomposes the HP subband into a number of directional subbands.

The NSPFB structure can avoid pseudo-Gibbs phenomena around singularities.

The NSDFB is built with two-channels NSFB, it is composed of iterated NSFB to decompose the NSFB into a number of directions, NSDFB structure can detect data in fine directions and represent the complex directional structure well. Looking for those properties, the NSCT is effective in representing well the detailed features of the strong dorsal hand vein texture along the radial and angular directions [17–19].

The NSCT coefficients are used as dorsal hand vein characteristics in the proposed dorsal hand vein identification method. For more details information on the NSCT, see reference [16].

3.3   Dorsal Hand Vein Features Extraction Using the NSCT

In order to extract the most discriminating characteristics in the dorsal hand vein with has a complex structure of texture, the NSCT which provides an optimum representation of image in space and spatial frequency is used in the proposed dorsal hand vein identification method.

We considered only phase information of NSCT decomposition outputs by assigning 0 or 1 to each bit of dorsal hand vein code according to the sign of the coefficients; 0 if the coefficient is negative and 1 if the coefficient is positive. So, the process of our approach is as following: The input image is first of all convolved with a NonSubsampled pyramidal filter Bank (NSP) which returns the details in the dorsal hand vein image at a single scale as shown in Fig. 4. This operation allows only the robust information in a dorsal hand vein to be passed on to the subsequent directional decomposition stage.

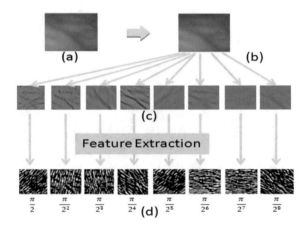

Fig. 4. Feature extraction: (a) Input image, (b) result of NSP decomposition: HP subbund, (c) the result of NSCT decomposition: directional subband, (d) result of phase response information which represent the dominant points existing in the each directional subband.

The direction of a dorsal hand vein feature decomposition is determined by the coefficient corresponding to the phase response among all directional subbands at a specific point. It's proven in [20] that the phase is more informative than the magnitude. We note that Cdi represent the coefficient at point (x; y) in the ith directional subband where i varied from 1 to 2k.

In a dorsal hand, vein patterns appear as dark intensities and correspond to a filter response. Phase response fixes well the orientation of a feature and corresponding to coefficient among all directional subbands.

The Phase Response Coefficients (Ph) of NSCT decomposition are defined as:

$$Ph \triangleq \angle(Cdi). \tag{1}$$

The Phase Response is computed in all subbands of image for the direction at location (x; y), the function assigns it to the ith direction is:

$$Ph_i^j = \left\{ \begin{array}{l} 1 \leq j \leq J \\ 1 \leq i \leq d_j \end{array} \right\}. \tag{2}$$

(J) represents the total number of scale decomposition and (dj) represents the number of directions at j-th scales. Ph_i^j represents the coefficients in the i-th directional subband of the j-th scale level. The coefficients Ph_i^j in each subbands at scale j are used like the dorsal hand vein features.

As is proved, this process will produce a large number of coefficients. If these coefficients are performed to the identification system directly, the speed of dorsal hand vein identification will be decreased, so we propose dividing the image into non-overlapping blocks as shown is Fig. 5 and then statistical feature is computed within each block. The feature which gives good results is the average and as result we have obtained a vector which size varies depending on the block size.

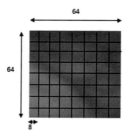

Fig. 5. Dorsal hand vein image blocks

By this way, the dimension of vector feature will be extremely decreased.

Then, the signs of statistical feature in each subband are used to generate the binary code (Bcod) which considered all the directional characteristics in both multi-directions and multi-scale:

$$Bcod_i^j(x, y) = \begin{cases} 1, & \text{if } mean\left(Ph_i^j(x, y) > 0\right) \\ & 1 \le i \le d_j \\ 0, & else \end{cases} \tag{3}$$

Denote that (x, y) represents the coordinates in each phase response of NonSub-sampled subband Ph_i^j. It is known that the resulting binary code (Bcod) contains sign information in each subband. The final binary vector of dorsal hand vein (Vcod) which contains all binary code resulting by the last step is expressed as:

And T represents the matrix transpose.

$$Vcod(x, y) = \Big[Bcod_1^1(x, y) \cdots Bcod_{d_1}^1(x, y) \vdots Bcod_1^2(x, y) \cdots Bcod_{d_2}^2(x, y) \vdots \cdots \vdots$$
$$Bcod_1^J(x, y) \cdots Bcod_{d_J}^J(x, y)\Big]^T. \tag{4}$$

3.4 Dorsal Hand Vein Feature Matching

After the characterization step, dorsal hand vein feature matching is presented to get the best decision of identification. For that, we use the Hamming distance (HD) in order to match between the obtained dorsal hand vein feature vectors. The HD is recommended here, because in the characterization we were used a binary code (with 0 and 1). The Hamming distance measured between two dorsal hand vein feature vectors Vcod1 and Vcod2 can be defined as [22]:

$$HD = \frac{1}{XY} \sum_{x=1;y=1}^{XY} Vcod^1(x, y) \oplus Vcod^2(x, y). \tag{5}$$

Where the pixel coordinates in the $X \times Y$ subband image is represented by (x, y). Thus, using the HD measure, we can determine whether the two dorsal hand vein vectors are generated from the same subject or not. So, if the value of the calculated HD is closer to '0', it means that the two dorsal hand vein vectors come from the same subject, and vice versa.

4   Experimental Results and Discussion

4.1   Database

In order to evaluated the performance of the proposed method, extensive experiments on GPDSvenasCCD dorsal hand vein image [13] are performed. Currently, GPDSvenasCCD presents the largest dorsal hand vein database available in public domain. The database consists of 10 different acquisitions of 102 people. Images were acquired in two separated sessions in one week: 5 samples the first time and other 5 samples the second session. The system to capture near infrared images of the hand dorsum consists of two arrays of 64 LEDs in the band of 850 nm, a CCD gigabit Ethernet PULNIX TM3275 camera with a high pass IR filter with 750 nm as cut off frequency (Table 1).

Table 1. Overview of the gpdsvenasccd database

Database	GPDSvenasCCD
Sensor	Non-contact
Person	102
Samples per identity	10
Total dorsal veins	1020

4.2   System Evaluation

The experiments are completed in identification mode (one-to-many). In order to prove the efficiency of the proposed encoding algorithm, we have calculated the rank one recognition rate. Then comparison of our method with curvelet transform method will be held. Experiments results are detailed in the Section below.

The methodology we have followed consists on randomly selecting one of dorsal hand veins images per subject form the gallery and treats all the remaining images as probes. The testing set is composed of 255 samples and the training set is composed of 765 in GPDSvenasCCD database.

We report cumulative Match Characteristic curve "CMC" (Fig. 6). This system achieved 99.96% of rank one identification rate for GPDSvenasCCD database.

To evaluate NSCT method, number of experiments are performed with different parameters as the number of decomposition level, the size of local region.

The orientation is ciphered by 2^k and k represents the number of decomposition level. Then, to choose the best number of decomposition k, we have varied k from 1 to 4 and for each stage we have calculate rank one recognition. The value of k = 3 (2^3 = 8 directions) achieves the best value of one rank recognition rate. For our approach in all

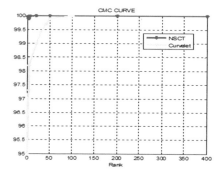

Fig. 6. Comparison of CMC curves with curvelet technique.

experiments and to reduce the size vector, we crop each ROI in blocks of (8 * 8) and (16 * 16) and to choose the best value, we have studied the influence of each image block extracted from filtered image and we calculate statistical descriptor in each block and we analyse our encoding to find the optimal block size. Experimental results indicate that computing NSCT with considering non-overlapping blocks of 8 * 8 achieve high performance recognition tests achieve 99.96% of rank-one recognition rate.

We compare the performance of the proposed method with Curvelet approach [10] which gives 95.02% of rank one identification rate Vs our method achieved 99.96% of rank one identification rate, those tests are applied on GPDSvenasCCD database using MATLAB implementation is up to 2.6 GHz, machine with 8 GB RAM.

5   Conclusion

We propose in this paper a dorsal hand vein Identification method, in which shift-invariant, multi-scale, and multi-directional NSCT coefficients are used as effective dorsal hand vein features. The first step is to fix the region of interest ROI in the same position of different dorsal hand vein images.

Next, the modified dorsal hand vein image is used as the input of the NSCT, then we applied phase response information of NSCT coefficients in each directional subband which are used to extract dorsal hand vein features. The decomposition gives a long feature vector so we proposed the variations of statistical feature computed in local region to decrease the size of the created dorsal hand vein vector feature to the size of 512 bytes. Finally, hamming distance is used for comparison and decision.

Experimental results show the effectiveness of the proposed NSCT feature based method in identification mode. We obtain an excellent result, of a 99.96% of rank-one recognition rate.

Future research will be focused on the performance of the proposed algorithm fusion with the multimodality of dorsal hand vein and palm vein images.

References

1. Lin, C., Fan, K.: Biometric verification using thermal images of palm-dorsa vein patterns. IEEE Trans. Circ. Syst. Video Technol. **14**(2), 199–213 (2004)
2. Cross, J., Smith, C.: Thermographic imaging of subcutaneous vascular network of the back of the hand for biometric identification. In: IEEE 29th Annual 1995 International Carnahan Conference, pp. 20–35 (1995)
3. Deepika, C., Kandaswamy, A.: An algorithm for improved accuracy in unimodal biometric systems through fusion of multiple feature sets. ICGST-GVIP J. **9**(3), 33–40 (2009). ISSN 1687-398X
4. Wang, L., Leedham, G.: Near-and-far-infrared imaging for vein pattern biometrics. In: Proceedings of the IEEE International Conference on Video and Signal Based Surveillance (2006)
5. Badawi, A.: Hand vein biometric verification prototype: a testing performance and patterns similarity. In: Proceedings of the 2006 International Conference on Image Processing, Computer Vision, and Pattern Recognition, IPCV 2006, Las Vegas, USA (2006)
6. Sanchit, Ramalho, M.: Biometric identification through palm and dorsal hand vein patterns. Instituto de Telecomunicaçées Lisbon, Portugal
7. Redhouane, L., et al.: Dorsal hand vein pattern feature extraction with wavelet transforms (2014)
8. Naidile, S., Shrividya, G.: Personal recognition based on dorsal hand vein pattern. Int. J. Innov. Res. Sci. Eng. Technol. **4**(5) (2015)
9. Jia, X., Cui, J., Xue, D., Pan, F.: An adaptive dorsal hand vein recognition algorithm based on optimized HMM. J. Comput. Inf. Syst. **8**(1), 313–322 (2012)
10. Ricardo, J., Augusto, F., Brandao, J.: A low cost system for dorsal hand vein patterns recognition using curvelets. In: First International Conference on Systems Informatics, Modelling and Simulation. IEEE (2014). 978-0-7695-5198-2/14 $31.00 © 2014
11. Miura, N., Nagasaka, A., Miyatake, T.: Extraction of finger-vein patterns using maximum curvature points in image profiles. Proc. IEICE – Trans. Inf. Syst. **90**(8), 1185–1194 (2007)
12. Rajarajeswari, M., Ashwin, G.: Dorsal hand vein authentication using FireFly algorithm and knuckle tip extraction. Int. J. Adv. Comput. Technol. (2014)
13. Ferrer, M.A., Morales, A., Ortega, A.: Infrared hand dorsum images for identification. Electron. Lett. **45**(6), 306–308 (2009)
14. Sonka, M., Hlavac, V., Boyle, R.: Image Processing, Analysis, and Machine Vision, 2nd edn., p. 26. PWS, New York (1999)
15. Oueslati, A., Feddaoui, N., Hamrouni, K.: Identity verification through dorsal hand vein texture based on NSCT coefficients. In: ACS/IEEE International Conference on Computer Systems and Applications AICCSA, Tunisia, November 2017
16. Do, M.N., Vetterli, M.: The contourlet transform: an efficient directional multiresolution image representation. IEEE Trans. Image Process. **14**(12), 2091–2106 (2005)
17. Tang, L., Zhao, F., Zhao, Z.G.: The nonsubsampled contourlet transform for image fusion. In: Proceedings of International Conference Wavelet Analysis and Pattern Recognition, Beijing, China, November 2007
18. Yang, B., Li, S.T., Sun, F.M.: Image fusion using nonsubsampled contourlet transform. In: Proceedings of Fourth International Conference on Image and Graphics, SiChuan, China, August 2007

19. Zhou, Y., Wang, J.: Image denoising based on the symmetric normal inverse Gaussian model and NSCT. IET Image Process. **6**(8), 1136–1147 (2012)
20. Oppenheim, A.V., Lim, J.S.: The importance of phase in signals. Proc. IEEE **69**, 529–541 (1981)
21. Zhou, Y., Wang, J.: Image denoising based on the symmetric normal inverse Gaussian model and NSCT. IET Image Process. **6**(8), 1136–1147 (2012)
22. Li, K.: Biometric Person Identification Using Near-infrared Handdorsa Vein Images (2013)

Author Index

Printed in the United States
By Bookmasters